Pulse crops for sustainable farms in sub-Saharan Africa

by

SIEGLINDE SNAPP
Department of Plant, Soil and Microbial Sciences
Michigan State University
East Lansing, United States of America

MARYAM RAHMANIAN
Plant Production and Protection Division (AGP)
Food and Agriculture Organization of the United Nations
Rome, Italy

and

CATERINA BATELLO
Plant Production and Protection Division (AGP)
Food and Agriculture Organization of the United Nations
Rome, Italy

Edited by

TEODARDO CALLES
Plant Production and Protection Division (AGP)
Food and Agriculture Organization of the United Nations
Rome, Italy

FOOD AND AGRICULTURE ORGANIZATION OF THE UNITED NATIONS
Rome, 2018

Recommended citation: Snapp, S., Rahmanian, M., Batello, C. 2018. *Pulse crops for sustainable farms in sub-Saharan Africa*, edited by T. Calles. Rome, FAO.

The designations employed and the presentation of material in this information product do not imply the expression of any opinion whatsoever on the part of the Food and Agriculture Organization of the United Nations (FAO) concerning the legal or development status of any country, territory, city or area or of its authorities, or concerning the delimitation of its frontiers or boundaries. The mention of specific companies or products of manufacturers, whether or not these have been patented, does not imply that these have been endorsed or recommended by FAO in preference to others of a similar nature that are not mentioned.

The views expressed in this information product are those of the author(s) and do not necessarily reflect the views or policies of FAO.

ISBN: 978-92-5-130088-6

© FAO, 2018

FAO encourages the use, reproduction and dissemination of material in this information product. Except where otherwise indicated, material may be copied, downloaded and printed for private study, research and teaching purposes, or for use in non-commercial products or services, provided that appropriate acknowledgement of FAO as the source and copyright holder is given and that FAO's endorsement of users' views, products or services is not implied in any way.

All requests for translation and adaptation rights, and for resale and other commercial use rights, should be made via www.fao.org/contact-us/licence-request or addressed to copyright@fao.org

FAO information products are available on the FAO Website (www.fao.org/publications) and can be purchased through publications-sales@fao.org.

Cover photographs (front and back): © Sieg Snapp

Contents

Foreword	v
Acknowledgements	vii
Abbreviations and acronyms	viii
Summary	ix

1. Introduction   **1**
    1.1 Pulses for sustainable agriculture and agroecology   1
    1.2 Pulses for family nutrition   3
    1.3 Pulses for income   5

2. Pulse cultivation in Africa   **5**
    2.1 Pulse classification   5
    2.2 Pulse cultivation in Africa   8
    2.3 Determinants of pulse and legume cultivation   10
    2.4 Pulse/legume growth type and services provided   11

3. Pulse/legume options for farming system niches   **12**
    3.1 Overview of farming system niches   13
    3.2 Genetic options   15
        3.2.1 Warm to hot semi-arid and sub-humid tropics   15
        3.2.2 Cool sub-humid to humid tropics   18
        3.2.3 Highland tropics   19
        3.2.4 Humid tropics   20
    3.3 Management options   20
        3.3.1 Semi-arid tropics   20
        3.3.2 Sub-humid to humid tropics   21
    3.5 Storage and seed technology options   22

4. How to promote pulses and legumes   **24**
    4.1 Seed systems   24
    4.2 Extension to promote pulses/legumes   26
    4.3 Extension of crop management   27

5. Research priorities   **28**
    5.1 Research priorities for pulses of the semi-arid and sub-humid tropics   31
        5.1.1 Common bean (*Phaseolus vulgaris* L.)   31

       5.1.2 Cowpea [*Vigna unguiculata* (L.) Walp.]   33
       5.1.3 Groundnut (*Arachis hypogaea* L.)   33
       5.1.4. Pigeonpea [*Cajanus cajan* (L.) Huth]   34
  5.2 Research priorities for pulses for the humid tropics   35
  5.3 Research priorities for pulses for the highlands   36
  5.4 Research priorities for pulses for irrigated
     rice-based systems   36
  5.5 New directions in research   36

6. Conclusions and recommendations   37

7. Key recommendations   37

References   38

Foreword

Pulses have a long history as staple crops for smallholders in sub-Saharan Africa. Cowpea [*Vigna unguiculata* (L.) Walp.], for example, has been used in Africa for millennia and is currently considered the single most important pulse in the dry areas of tropical Africa. One of the oldest crops ever farmed, cowpea is known to thrive in challenging conditions – sandy soil and scarce rainfall.

Pulses, and legumes in general, have also played an important role in soil health maintenance and improvement of Africa's nutrient-poor soils. However, over time, consumers' preferences have changed, giving way to important cereals (e.g., rice, cassava, maize), subject of vast research and political support worldwide. The quantity of arable land used for pulses is much less than the area cultivated with important cereals, thus negatively affecting the nutrient balance in African soils. Pulses play a role in improving soil fertility due to their ability to biologically fix atmospheric nitrogen and, some of them, enhance the biological turnover of phosphorous.

Due to their rich nutritional value, pulses are an important part of a balanced, healthy diet. Pulses are a good source of protein and of micronutrients such as iron and zinc. Pulses can play a key role in fighting iron deficiency anaemia, one of the most important micronutrient deficiencies in sub-Saharan Africa, and protein and energy deficiencies, in both quantity and quality, as they are often the cause for widespread malnutrition, which manifests itself in the form of stunting or wasting. Moreover, pulses are low in fat and high in dietary fibres, which among others slows the absorption of lipids and lowers blood cholesterol levels as well as helping with digestion. Nutritional benefits of pulses can play an important role in reducing hunger and increasing human health, so contributing towards the achievement of United Nations (UN) Sustainable Development Goals. However, these benefits are often underappreciated and therefore the UN General Assembly declared 2016 as the International Year of Pulses (IYP), which sought to recognize the contribution that pulses make to sustainable agriculture, human well-being and the environment.

FAO's Regional Office for Africa in Ghana and FAO's Plant Production and Protection Division at Rome Headquarters support countries to sustainably intensify agricultural production in Africa, and pulses are often part of that agenda. Among the aims of the regional initiative, "Sustainable production intensification and value chain development in Africa" is the enhancement of agricultural diversification and promotion of innovative cropping practices. In this context, pulses could be integrated into cassava, maize and rice production systems, which are very important crops as they provide around 40 percent of Africa's food. The stronger integration of pulses help to sustainably intensify these production systems and value chains could be further developed.

There have been considerable research efforts to develop strategies to support pulses cultivation and utilization on smallholder farms in sub-Saharan Africa, which has resulted in a large body of published and unpublished data. However, an authoritative review in this area has been lacking. To fill this gap and to raise awareness on the importance of pulses in sub-Saharan Africa, authors have collated and synthesized the available information in this comprehensive state-of-the-art document. It highlights strategies with a high potential to improve existing key cropping systems in sub-Saharan Africa.

This review improves our current knowledge on pulses and associated technologies in sub-Saharan Africa and shall also motivate to increase the utilization of pulses in crop production. It is a useful reference for researchers, extension workers, policy-makers and donors alike.

Bukar Tijani
Assistant Director-General
Regional Representative for Africa
Regional Office for Africa
Food and Agriculture Organization
of the United Nations

Hans Dreyer
Director
Plant Production and Protection Division
Food and Agriculture Organization
of the United Nations

Acknowledgements

Our sincere thanks to all the visionaries who work with farmers, who are farmers, and through their dedication are developing sustainable agriculture and food systems, one community at a time. Their invaluable insights, and commitment are the inspiration for this review. The main author gives her sincere appreciation to Teodardo Calles, who has the special charge for legumes among the many staff that have worked so hard on making a success of the International Year of the Pulse. This is but one example of the unceasing support from all the scientists and staff at Food and Agriculture Organizations of the United Nations. The editor would like to express his gratitutde to Claudia Nicolai, Diana Gutiérrez, Michela Baratelli, Ricardo del Castello, Maria Xipsiti and many others who have helped in preparing this publication. The authors and editor would like to thank the three anonymous peer reviewers, who made important suggestions for improving the final manuscript. The main author would like to thank her family members for their support and understanding and for making this review possible.

The authors would like to thank Thorgeir Lawrence for final editing to conform to FAO editorial style, Chrissi Smith for laying out the document and Suzanne Redfern for designing the cover.

Abbreviations and acronyms

BNF	Biological nitrogen fixation
C	Carbon
CGIAR	Consultative Group on International Agricultural Research
DLS	Doubled-up legume system
FAO	Food and Agriculture Organization of the United Nations
FFS	Farmers' field school
IPM	Integrated pest management
MAC	Mid-altitude climbers
masl	Metres above sea level
N	Nitrogen
NGO	Non-governmental organization
P	Phosphorus
PABRA	Pan-Africa Bean Research Alliance
PICS	Purdue Improved Crop Storage (bags)
QD	Quality declared
SSA	sub-Saharan Africa

Summary

Food insecurity in sub-Saharan Africa is a problem affecting 153 million individuals (ca. 25%). This problem could be worsen by the ongoing soil degradation, being cause by the reduction of soil organic matter and insufficient nutrient supply. Over 75% of the agricultural land in Africa could be classified as degraded by 2020. This situation can compromise food production in sub-Saharan Africa, both quantitatively and qualitatively, and the sustainability of existing agricultural production systems.

The use of fertilizer could revert this situation; however, Africa has almost no capacity to produce fertilizers (African fertilizers production facilities work mainly in blending fertilizers) and therefore fertilizers are produced elsewhere outside Africa and transported from long distances at great expenses. This situation grants to sub-Saharan Africa farmers only a very limited access to fertilizers, thus increasing the risk of soil degradation.

Pulses have a long history in sub-Saharan Africa due to their multiple benefits. Pulses, and legumes in general, can play an important role in agriculture because their ability to biologically fix atmospheric nitrogen and to enhance the biological turnover of phosphorous; thus they could become the cornerstone of sustainable agriculture in Africa. In this sense, there is a body of literature that points to diversification of existing production systems; particularly legumes species, which provides critical environmental services, including soil erosion control and soil nutrient recapitalization. This publication is a review of some of the promising strategies to support pulses cultivation and utilization on smallholder farms in sub-Saharan Africa. The review is part of the legacy of the International Year of Pulses (IYP), which sought to recognize the contribution that pulses make to human well-being and the environment.

One challenge faced worldwide is that the diversity of pulses are not captured well in statistics. There is not a clear picture of what is grown and where, and this leads to an under-estimation of their importance for sub-Saharan Africa and consequently reduce research investment in pulses. Existing agricultural production systems are dominated by cereals, and represent opportunities for enhanced crop diversification, through promoting local and novel pulse varieties. Mixed-maize is a system that is rapidly growing and poses one such opportunity, particularly for beans. This is due in part to the large number of bean varieties that have been developed to meet local and regional market requirements, through decades long partnerships foster by Pan-Africa Bean Research Alliance (PABRA). Bean research has included pioneering participatory plant breeding, extension linked to participatory community organizations and value chains, as well as attention to informal seed systems. This example shows how pulse research can make a different on smallholder farms in sub-Saharan Africa, by broadening the range of genetic options and supporting innovation. There are many such farmer-approved varieties available that deserve greater promotion, as do technologies such as doubled up legume system innovation recently released by the Malawi government. At the same time, this review has highlighted that variety release has lagged for some pulse crops, and that there is urgent need for more research on adoption, barriers to adoption, and on impact of adoption.

Research priorities suggested include greater recognition and attention to expanding properties associated with *multipurpose* types of pulses, which are popular in sub-Saharan Africa. Different types of pulses are needed for different functions and in general, *multipurpose* pulses are the best to respond to the diverse needs of farmers, including food, fuel and fodder, and ecosystem services such as pollination. There is a trade-off between the harvest index and other functions, which have too often been overlooked by researchers and decision makers who tend to focus almost exclusively on increasing grain yields. Pest tolerance, as well as extension of educational approaches and agronomic advice to strengthen integrated pest management (IPM) is another area urgently needing attention. Finally, the role of specific legumes and associated biochemical properties in promoting ecosystem health, community health – this is a crucial area for research that will provide urgently needed options for women farmers – and for sustainability of communities.

1. Introduction

The United Nations declared 2016 to be the International Year of Pulses, which provided a unique opportunity to review and consider how to strengthen key contributions made by pulse crops. A pulse is defined here as a leguminous crop that produces a dry grain used for food and feed (FAO, 1994). Legumes belong to the Leguminosae family, which is the third-largest flowering plant family, and comprises species with unique attributes that make essential contributions to sustainable livelihoods, nutrition and sustainable production systems. Soil fertility maintenance is supported worldwide by the presence of legumes, as is the production of diverse, protein-rich plant products and livestock fodder.

This document reports on a literature review that assessed promising pulse options to safeguard food security, nutrition and environment conservation on African smallholder farms. Our primary target here is the five legume crops grown most widely in sub-Saharan Africa (SSA), namely common bean (*Phaseolus vulgaris* L.), chickpea (*Cicer arietinum* L.), cowpea [*Vigna unguiculata* (L.) Walp.], groundnut (*Arachis hypogaea* L.), and pigeonpea [*Cajanus cajan* (L.) Huth]. Note that in addition to the pulse species, we have included groundnut in this review, even though it is not considered a pulse. We chose to include it based on its widespread production across Africa [almost 14 million tonnes produced according to FAO (2016)], where it is the second most important grain legume, and a key source of plant-protein, oil and income for many smallholder farmers. A similar rationale for the choice of pulses plus groundnut is presented in a recent FAO review of regional legume trade (Koroma *et al.*, 2016).

Agroecological zones are used as a structure for this review. The identification of legume crops and management systems was carried out in relationship to farming system niches, from the arid tropics to the humid tropics. Major extension and research opportunities and gaps were also identified using this framework. Finally, we describe recommended options for promoting use of legume crops and adjusting government policies.

1.1 PULSES FOR SUSTAINABLE AGRICULTURE AND AGROECOLOGY

Integrated farming systems that rely primarily on biological processes are fundamental for agro-ecologically-based food production (Marinus *et al.*, 2016). The presence of legumes is the only alternative to synthetic fertilizers, which require fossil fuels to support their production through the Haber Bosch process for nitrogenous fertilizers, and through mining of scarce resources for phosphorus (P) fertilizers (Drinkwater and Snapp, 2008). There are serious environmental concerns associated with fertilizer use, including increased occurrence of invasive plant species, impaired water quality, and contributions to greenhouse gas generation, as well as 'dead zones' in large bodies of water (Rockström *et al.*, 2016). It is also important to note that fertilizers are very expensive for farmers in Africa, a continent with almost no capacity to produce fertilizers. African fertilizer production facilities are limited to blending fertilizers, which are produced and transported at great expense from long distances.

Soil degradation is a global problem, with insufficient nutrient supply and soil organic matter declines threatening food production in Africa, and beyond (Bekunda, Sanginga and Woomer, 2010). Over three-quarters of agricultural area in Africa will be classified as degraded by 2020, according to one review (Scherr, 1999). There is a body of literature that points to diversification, and particularly legumes as a plant species group that provides critical environmental services, including soil erosion control and soil nutrient recapitalization (Fornara and Tilman, 2008; Powell, Pearson and Hiernaux, 2004). Organic matter technologies for integrated fertilizer management are both directly and indirectly dependent on legume presence, as manure is an important source of organic amendments, yet the quality and quantity of manure depends on livestock feed, and quality of feeds depends on legumes (Snapp, Mafongoya and Waddington, 1998). A strong case can be made that sustainable agriculture requires that a substantial proportion of crop land be dedicated to growing legumes (Barrios, Buresh and Sprent, 1996; Koutika *et al.*, 2005). This often involves polycultures or mixed cropping systems that include legumes. However, the life history and growth type of the legume also matters. For example, a legume that is highly vegetative in growth habit and that has a long-duration growth period will be highly effective at conserving soil and at enhancing soil fertility. Whereas with a short duration, early maturing, high harvest index type of legume, such as dwarf bean, the growth type will not provide much soil benefit. It can be argued that the greater the longevity of a legume, the greater the contribution to soil rehabilitation; this is noted in a recent call for more perennial solutions to agricultural challenges in Africa (Glover, Reganold and Cox, 2012). At the same time, the environmental and socio-economic context will determine what proportion of legumes are grown, and what growth type of legume is preferred by farmers. The context will also influence what proportion of legumes will be needed to support environmental services.

Legumes are vital for sustainable production because they directly supply biologically fixed nitrogen, and they are also the indirect source of manure-based nitrogen inputs to maintain soil productivity (Giller and Cadisch, 1995). Although not all legume species form symbiotic relationships with Rhizobia, it is an important trait associated with all cultivated legumes. If a root system is well-nodulated, with effective Rhizobia, then the amount of nitrogen fixed is generally about 30 to 40 kg of N per tonne of biomass produced (Peoples *et al.*, 2009). There are many factors that influence the amount of biologically fixed nitrogen, but the growth of the plant is an overriding one (once a successful symbiosis is achieved). Inoculation of legumes with appropriate Rhizobia species is an important agricultural practice, particularly in cases where there are insufficient or non-beneficial indigenous Rhizobia. It depends on the species, as legume crops that are indigenous to an area are often not responsive to inoculation. Cowpea is such an example in Africa.

Legumes are a primary source of high quality feed that is critical for sustainable intensification of livestock systems. This includes dairy value chains, and confined feeding. Further, a legume presence is a crucial component of any effort to diversify and enrich the quality of feed in rangeland or pastures. Extensification rather than intensification is generally characteristic of livestock production in Africa, and there has been considerable discussion in the literature regarding how feasible fodder plantings are in an African smallholder context. Africa farmers generally follow extensive agricultural practices in contrast to the intensive nature of forage production, and indeed it has been argued that very little to nil adoption of forage legumes has occurred to date (Sumberg, 2002).

There are barriers to adoption of agricultural practices that involve investment in plant production specifically for animal feed, namely the requirement for sufficient income and returns from livestock to support such investment; this often involves a dairy value chain, or high-value meat production (Tarawali et al., 1999). There are however further challenges, beyond developing profitable technologies. Research is clearly needed to understand the barriers, and the socio-economic niches where forages and multipurpose crops could provide farmer-relevant opportunities. Important considerations need to be addressed, these include farmer preferences and targeting technologies through recommendation domains and/or better understanding of demand (Sumberg, 2002). Leguminous crop residues used as fodder has been put forward as one key technical option, as a means to improve livestock feed quality while still ensuring that farmers received adequate returns, based on dual use; legumes as a food or income crop; and as a fodder source (Powell, Pearson and Hiernaux, 2004). Recycling of nutrients from legumes through animal ingestion and manure management pathways also requires investment in knowledge, infrastructure and means of transportation. Thus, there are multiple steps involved in adoption of legumes, particularly if this practice is considered as part of a crop-livestock intensification and agricultural development process.

1.2 PULSES FOR FAMILY NUTRITION

One of the critically important sustainable development goals is to end all forms of malnutrition, and specifically to markedly reduce child stunting. One of the most pernicious forms of malnutrition is caused by insufficient protein, a complaint widespread in Africa. Food legumes are in particular valued for their nutrient-rich products, which include grain, and in many cases, vegetable in the form of leaves and pods (Dixon and Sumner, 2003). Animal feed also involves the high-protein plant products of legumes. This includes stems, leaves, as well as mature and immature pods and seeds. The nutritional benefits of legumes include not only high protein content, but also a unique amino acid complement that includes tryptophan and lysine (Asif et al., 2013). These amino acids are in short supply in cereals. Cereals are also deficient in zinc and folate, an important iron source, both of which are supplied by most legume products. Thus legume foods are highly complementary to cereals, the major source of calories in diets around the world.

Pulse crops provide needed dietary diversity, and thus are key contributors to human nutrition as a crucially important source of protein, of amino acid diversity and of B-group vitamins, iron, zinc, magnesium and calcium (Messina, 1999). Amino acid diversity is essential, as the combination of amino acids that make up protein associated with cereal crops and that of legume crops are at the foundation of a balanced diet around the world, particularly in locations where plant-based food products are relied upon for family nutrition. Phytochemicals and dietary fibre are other important nutritionally beneficial attributes associated with consuming a legume-rich diet.

The high nutrient content of legumes is closely associated with complex and unique biochemistry. Through evolution, many legumes have developed unique sets of biochemical compounds as defense mechanisms against the expressed preference by insects and other animals to eat nutrient-rich legume tissues, and in some cases as adaptation to stress in marginal environments (Cullis and Kunert, 2016). Thus it is not surprising that there are many antinutritional as well as nutritional properties accompanying legume seeds and food products. These include enzyme inhibitors, lectins, polyphenolics and in some cases tannins (Odeny, 2007). Such compounds are often associated with reduced nutritional value of food, as digestibility is reduced as well as bioavailability of some minerals. Treatment of legume grains through steps such as mechanical, biological (e.g. fermentation) or heat is often necessary to remove anti-nutritional properties (Deshpande et al., 1982) and improve bio-availability of some minerals. Plant breeding efforts are also important, and have been notably successful at developing pulse cultivars with high micronutrient content. At the same time, it is vital to pay attention to farmer preferences in conjunction with nutritional improvements. This is illustrated by the story of bean improvement in Africa. Research on Andean beans has proven to be essential to achieving success – where Andean germplasm is the source of the larger seed type and seed quality that is overwhelming preferred in SSA. Attention to this neglected germplasm was critical for successful introduction of beans with enriched levels of iron, zinc and phosphorus (Cichy et al., 2009).

Evidence of legume impacts on family nutrition is expected to be difficult to show, as health and nutrition are influenced by diet through many direct and indirect pathways, and consequently there have been few studies elucidating this important topic. One such was carried out in Uganda, where food security – but not income – was enhanced through the promotion of improved bean cultivars (Larochelle et al., 2015). Another study, in Northern Malawi, found that child stunting was reduced in villages where participatory research had been carried out, where the interventions combined nutrition education with legume crop diversification (Bezner-Kerr, Berti and Shumba, 2011). More research is clearly needed on such connections.

1.3 PULSES FOR INCOME
The role of grain legumes in African agricultural systems is multifaceted, and many pulses provide an important source of income as they can be sold for high prices at local or international markets. The price often reflects the nutrition-packed nature of legume grains, with a high protein content, as well as unique amino acid complements, and specific biochemical compounds that vary with species. At the same time, a household survey in Malawi indicated that farmers were not always realizing a profit from legume sales, as labour inputs were high, access to good seed was poor, and legume prices varied tremendously (Snapp *et al.*, 2002). Further, studies in India have documented that legume prices are often not regulated or supported by government policies, and vary tremendously from one market to the next, and over time – this is in contrast to the general stability associated with cereal prices (Rao, 2000). There is need for market research to assess the extent to which a price premium is associated with legume crops – relative to staple cereal crops – as well as volatility, and how this influences smallholder farmer incentives to grow legumes.

Post harvest handling together with storage and processing, are essential aspects of a legume grain value chain. All have important implications for income generation potential. Particular attention needs to be paid to education on these topics, and to infrastructure that supports smallholder involvement in the entire legume value chain. This particularly holds for women farmers. An example is presented in Box 1, which highlights the benefits – and challenges – associated with cowpea processed into a nutritious snack. In West Africa, this can be an important source of agricultural-based income.

2. Pulse cultivation in Africa

2.1 PULSE CLASSIFICATION
This section provides a brief introduction to classification of legume crops, and an overview of legume crops and utilization in Africa. This provides important context for the review.

The *Leguminosae* family is divided into three sub-families: *Caesalpinioideae*, *Mimosoideae* and *Papilionoideae* (Lewis *et al.*, 2005). According to this source, the sub-family *Papilionoideae* is the largest, comprising 28 tribes. This sub-family includes all species considered pulses, as well as other important grain legumes. As shown in Table 1, according to FAO (1994), pulses is the term

> **BOX 1.**
> **West Africa cowpea: opportunities and challenges**
>
> Cowpea is a 'golden grain' for many women entrepreneurs in West Africa, where it is used to produce a golden fried donut snack that is called *'kosai'*. It is a nutritious type of donut, complementing staple grains with a high quality protein that makes up about 25 percent of the cowpea seed (by weight). Income generation depends on the experience of the farmer-chef-marketer, and is influenced by local demand for *kosai*; however, returns are manyfold higher than minimum wage. Cowpea as a snack, and as a staple protein source, is in constant demand. West African production has increased over the last decade at about 10 percent annually, from Niger to Nigeria. New varieties of cowpea include a number of popular dual purpose types, that are critically important sources of high quality forage for livestock finishing purposes. This is another income opportunity, as fattening livestock for religious holidays is a relatively steady source of agricultural income in West Africa, amidst the uncertainty of rainfall and market fluctuations. There are many challenges associated with producing cowpea, as it is such a nutritious treat that pests such as aphids and pod borers plague it. Unsafe levels of pesticide residues in cowpea have been shown to be a problem, underlying the urgency for improved education on less-toxic pest control measures. There is clear need for the development of more pest resistant varieties and for promotion of non-toxic storage methods, such as 'PIC' bags.

for the dried grain product of beans (different *Phaseolus* and *Vigna* species), pea (*Pisum sativum* L.), pigeonpea, cowpea, chickpea, lentil (*Lens culinaris* Medik.), broad bean (*Vicia faba* L.), Lupins (different *Lupinus* L. species), vetch (*Vicia sativa* L.), Bambara bean [*Vigna subterranea* (L.) Verdc.] and pulses nes[1] (aggregated category including species of minor international relevance). Other legumes cultivated across Africa include soybean [*Glycine max* (L.) Merr.], groundnut and many forage species such as Brazilian lucerne [*Stylosanthes guianensis* (Aubl.) Sw.] and leucena [*Leucaena leucocephala* (Lam.) de Wit].

The focus of this review is on common bean, chickpea, cowpea, and pigeonpea, with the addition of groundnut, as these are the staple legume food crops grown on African smallholder farms. Legumes that are cultivated in Africa on a more occasional basis will also be considered, and we note that these are often locally important. See below for an expanded description. It is important to note that legumes are often grown for multiple purposes, including fodder, fuel wood, medicinal purposes and cash sales as well as food.

---

[1] Stands for "not elsewhere specified".

Table 1. Classification of pulses according to FAO (1994)

FAO code	Commodity	Remarks[1]
176	Beans, dry	This is aggregated category that includes the following species: 1) common bean (*Phaseolus vulgaris*), 2) lima bean (*Phaseolus lunatus*), 3) scarlet runner bean (*Phaseolus coccineus*), 4) tepary bean (*Phaseolus acutifolius*), 5) adzuki bean (*Vigna angularis*), 6) mung bean (*Vigna radiata*), 7) mungo bean (*Vigna mungo*), 8) rice bean (*Vigna umbellata*) and 9) moth bean (*Vigna aconitifolia*).
191	Chickpeas	This category only includes chickpea (*Cicer arietinum*).
187	Peas, dry	This category only includes pea (*Pisum sativum*).
195	Cowpeas, dry	This category only includes cowpea (*Vigna unguiculata*).
201	Lentils	This category only includes lentil (*Lens culinaris*).
197	Pigeon peas	This category only includes pigeon pea (*Cajanus cajan*).
181	Broad beans	This category only includes broad bean (*Vicia faba*).
210	Lupins	This category includes several species of the genus *Lupinus* L.
205	Vetches	This category only includes vetch (*Vicia sativa*).
203	Bambara beans	This category only includes Bambara beans (*Vigna subterranea*).
211	Pulses, nes[2]	This is aggregated which includes species of minor relevance at international level: 1) hyacinth bean (*Lablab purpureus*), 2) jack bean (*Canavalia ensiformis*), 3) winged bean (*Psophocarpus tetragonolobus*), 4) guar bean (*Cyamopsis tetragonoloba*), 5) velvet bean (*Mucuna pruriens*) and 6) African yam bean (*Sphenostylis stenocarpa*).

1. Scientific names are sourced from the updated taxonomic database Tropicos (MBG, 2016).
2. 'not elsewhere specified'.

Soybean is also a leguminous crop of growing importance across Africa, but this crop is rarely used for food, and has been reviewed elsewhere (Gasparri *et al.*, 2016); it was not included in this review.

In 1994, FAO published categories for reporting pulses, which have not changed in any substantive manner since (Table 1). This classification system includes eight clear cases, where species are reported as a unique species, and three aggregated classes ('bean, dry', 'lupins' and 'pulses nes'). This use of aggregated classes is a source of considerable confusion. For example, the dry beans combined category leads to the reporting of mung bean [*Vigna radiata* (L.) R. Wilczek] and common bean in one class, yet these crops are from two distinct genera, with quite different environmental and market niches. There is also additional confusion because some users of FAO statistics assume that the category 'beans, dry' only included *Phaseolus* species (as originally intended) or, what is more problematic, the widely cultivated common bean species (Deshpande *et al.*, 1982; Eitzinger *et al.*, 2016).

Nomenclatural confusion abounds, with many crops referred to as some type of 'bean' or a 'pea' including many that are not closely related, so that tracking which legumes are grown where is incomplete and inaccurate. This may have contributed to limited appreciation of the scope and diversity of legume crops that are utilized around the world. Indeed, the classification confusion may have led to under-reporting and, ultimately, under-investment in research, extension, and policy that takes account of legumes within agricultural production systems. The extent to which nomenclatural confusion is a symptom of lack of attention to this plant family, or is a cause, is hard to ascertain. Inadequate documentation of what legumes are grown and where; a real lack of adoption studies; and overall failure to invest in legume crop improvement, were some of the startling findings of a recent review of CGIAR research on legume crops (Pachico, 2014).

An important contributor to what appears to be under-representation of legumes in agricultural statistics, is that many species are grown as intercrops on smallholder farms around the globe. Mixed cropping systems and locally consumed crops are often ignored in favour of sole-cropped crops, and particularly those that are sold on export markets, i.e. the crops that are also commodities. Many tropical legumes are included in the aggregated categories of 'not elsewhere specified' or 'bean', and so are not documented individually. These include velvet bean [*Mucuna pruriens* (L.) DC.], hyacinth bean [*Lablab purpureus* (L.) Sweet], African yam bean [*Sphenostylis stenocarpa* (Hochst. ex A. Rich.) Harms], tropical lima bean (*Phaseolus lunatus* L.) and runner bean (*Phaseolus coccineus* L.). These so-called minor legumes are primarily grown on smallholder farms for home consumption and local markets. This may be a contributing factor to the above-noted lack of documentation and poor statistics for many legume crops. Taken together, these factors lead the authors of this review to suggest there has been consistent under-representation of minor legume crops in national agricultural production statistics, and, in turn, these crops have rarely been prioritized for research investment. Recent attention has included suggestions that genomic tools could be used to characterize and improve such crops, although a major investment of resources would be required as there has been very little research to date carried out on these crops (Cullis and Kunert, 2016).

In a landmark study, over two hundred tropical legumes were identified as deserving research attention for the role they might play in agricultural development (NAS, 1979). This present study focuses on only a subset of these species, those that historically were cultivated across much of Africa but have been neglected in recent years. It is recognized that there is a tremendous genetic resource base of tropical legume species that deserve plant breeder attention. There is therefore a clear call for research that is much broader in scope than current studies, which often focus on identification of novel resistance to biotic and abiotic stresses, for transfer to a few select crops. This is the case for tepary bean (*Phaseolus acutifolius* A. Gray) that is being investigated as a source of drought and heat tolerance for genetic improvement of common bean (Beebe *et al.*, 2011). There is urgent need for sustained and well-resourced efforts to improve minor tropical legume crops in their own right (Cullis and Kunert, 2016).

2.2 PULSE CULTIVATION IN AFRICA

The status of pulse production in Africa is one that is difficult to gauge; as noted earlier, global agricultural statistics are notably poor for legumes. This

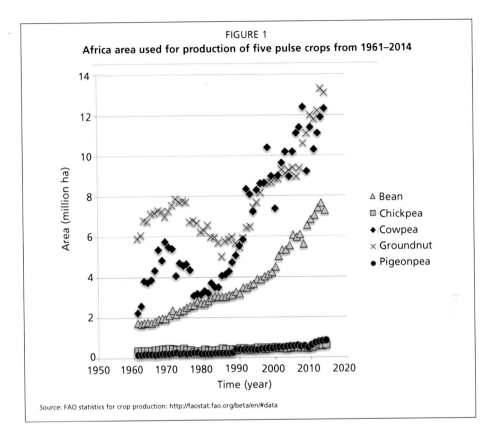

FIGURE 1
Africa area used for production of five pulse crops from 1961–2014

Source: FAO statistics for crop production: http://faostat.fao.org/beta/en/#data

is partly due to the tremendous diversity of legume species, where many are grown in mixed or intercropped systems that can lead to legume species being overlooked in favour of a cereal or root crop that is also present. There are also definitional issues in agricultural statistics, with many species being misidentified or reported in an aggregated manner, as some crop categories include a composite of quite different species. Indeed, there is considerable nomenclatural confusion at local as well as international levels; names that include 'bean' and 'pea' are often interchangeably applied to completely different species and reported erroneously. This problem occurs within the scientific literature, and throughout the legume production chain, from field to market place. Research is urgently needed that documents legume germplasm distribution – including which varieties are grown, and where; –this is a tremendous knowledge gap and lags behind that of other food crops (Pachico, 2014).

Overall, pulse consumption has declined in regions with high incomes, such as in the Americas and across Europe. In Africa, in contrast, the pulse production area continues to expand, at a rapid rate in some cases (Figure 1). Driving much of this expansion may be a 'pull' from market demand, at both the local and international level (Koroma et al., 2016). A notable example is the steady expansion in area dedicated to bean production. Cowpea and groundnut area both declined in the 1980s, then resumed and are rapidly increasing across

Africa today. Pigeonpea and chickpea are largely confined to specific areas in Southern and East Africa for pigeonpea, and Ethiopia for chickpea. The area cultivated with chickpea has been largely static, although yields have increased recently, particularly in Ethiopia (FAO, 2014). Pigeonpea has been grown for many years but on only a modest acreage in Africa (Figure 1). This may in part reflect the undercounting of a crop that is almost always grown as an intercrop with maize, or in some cases with sorghum or other crops. Crops grown as intercrops may be undercounted in agricultural statistics. At the same time, the area devoted to pigeonpea cultivation is increasing rapidly over large areas of Malawi, the United Republic of Tanzania and Mozambique (Walker et al., 2015). There is strong, site-specific demand for minor legume crop species, as shown by the high market price in Kenya for hyacinth bean (N. Miller, pers. comm., 2016), and new markets for velvet bean [*Mucuna pruriens* (L.) DC.] associated with goat auctions in Zimbabwe (Kee-Tui et al., 2015).

Pulse yields have not increased at nearly the rate of cropping system area expansion. This implies that the expansion of area planted to legume crops is only partially due to rising demand. The extensive farming systems that remain a dominant feature in Africa may be a major explanatory factor to rising legume crop area, as there are only localized areas of crop intensification. A mixed-maize-bean production system across the central-southern part of the United Republic of Tanzania and much of southern Africa is one such example of crop intensification systems where legume crop yields are rising, along with maize (Blackie and Dixon, 2016).

2.3 DETERMINANTS OF PULSE AND LEGUME CULTIVATION

Farmers often choose which crops to grow based on family food preferences, culture, food security needs and market opportunities. Other important influences on which legumes are grown include knowledge of production and post-production practices, and access to seed (Mhango, Snapp and Kanyama-Phiri, 2013). There are a number of barriers to legume production, including a focus on cereals and cash crops that often ignores legumes, as shown by many government policy and research institution priorities (Isaacs et al., 2016b). The urgent requirement to meet calorie goals on small acreages goes some way to explain the attention often given to cereals.

A market value chain approach considers the entire chain, providing support for improvement of technologies, knowledge and infrastructure from seed through to grain production, including post-harvest processing and marketing (Steele, 2011). Such approaches have helped farmers and farmer groups to link to opportunities to sell legumes for high value direct consumption (e.g. groundnut), or for storage, processing and links to local, regional and global markets (e.g. common bean and pigeonpea, Kaoneka et al., 2016). How to support expansion of informal and formal seed systems, and extension approaches are discussed below under means to promote legume cultivation in Africa.

2.4 PULSE/LEGUME GROWTH TYPE AND SERVICES PROVIDED

The growth type of the legume influences the types of services it provides. Crop legumes vary from short-lived plants with lifecycles of two to three months, and other types of plants that live for two or more years. If a crop is short-lived, and produces high grain yields, this type of legume helps with food security but does not enhance soil nitrogen status (Giller and Cadisch, 1995). A very different type of legume has a long life cycle, growing for 6, 9 or more months, and producing large amounts of biomass for multiple benefits. Such plants develop deep root systems and a large symbiotic apparatus that supports biological nitrogen fixation (BNF) and P mobilization over many months, and builds soil fertility. This type of long-lived legume often produces lots of biomass aboveground for fodder, as well as food, income and soil fertility benefits.

A number of legume species have an unusual ability to enhance solubilization of P, an important mechanism for the enhancement of soil fertility (Drinkwater and Snapp, 2008). Legume root exudates enhance the biological turnover of P in specific soil types, moving P from "sparingly available" to "soluble" pools that are available to other plants. Such mechanisms are species dependent and include many tropical food legumes, including chickpea, groundnut and pigeonpea (Richardson et al., 2011).

Residue management is a largely unknown and unconsidered aspect that is none the less an important factor in legume contributions to global agriculture. Modelling the terrestrial carbon budget requires attention to agricultural practices, and several studies have highlighted the lack of certainty occasioned by major unknowns, such as the extent and intensity of crop residue management (Bondeau et al., 2007). There are also tradeoffs to consider, as in general agro-ecozones that support high net primary production of crops, with associated residues. These are also the areas where utilization of crop residues for livestock and other uses (fuel for example) leaves little residue behind for soil improvement purposes. This limits soil benefits from residue produced in high potential, mesic rainfall environments, whereas in low potential, marginal environments where residues are rarely collected, there is little to utilize, as primary growth of crops and residues is poor (Valbuena et al., 2012).

There are numerous services that legumes can provide, and some key benefits have been outlined here. The relationship of growth type to services provided has been emphasized along with the importance of management to maximize returns from legume integration. However, adoption of legumes will depend to a large degree on the context, including the socio-economic environment and biophysical conditions. This review next considers which legumes fit within which niches, as context is a key conditioning criteria for selection of legumes to promote the potential for their successful adoption.

3. Pulse/legume options for farming system niches

Introducing a wider range of varieties is fundamental in efforts to support farmer experimentation and to promote legume production. Modern cultivars and traditional landraces can both be important sources of diversity. Through promoting participatory on-farm testing of a wide diversity of varieties, legumes with new or missing traits can be introduced to a community or a farm family, for purposes such as disease resistance and tolerance to poor soil types. This exposure to a wide range of germplasm will help diversify farms, particularly if accompanied by education on the value of diversity and on agroecology. However, not all legumes are expected to thrive in all environments. This is why we have taken the approach of considering the biophysical and socio-economic niche in relationship to legume varieties and management systems that are promising options for dissemination through extension efforts (Table 2).

Identification of legume options with a good chance of success is greatly enhanced by first taking into consideration the farming system environment, through the concept of a socio-ecological niche (Ojiem et al., 2006). Social and economic context as well as biophysical properties are thus used to define a niche to predict which legume technologies might be a good fit. As a first step in identification of promising pulse options it would be useful to consider

Table 2. Farming system ecological niches and associated legume options that are proven and ready for dissemination through extension. More detail is provided in Table 3.

Niche	Genetics	Agronomy system	Agronomy management	Post-harvest
Arid zone	Local cowpeaPlus[1] varieties, adapted to harsh environment, for food and fodder	Cowpea-millet and cowpea-sorghum		
Cereal semi-arid	Early Cowpea[2] CowpeaPlus	Relay shade-tolerant cowpea Groundnut disease resistant varieties	Seed priming	FFS Cowpea seed systems Cowpea – PIC bags
Mixed-maize	Early Bean PigeonpeaPlus	Intercrop with shade-tolerant bean Ratoon pigeonpea Pigeonpea-groundnut DLS[3]	Cereal-legume mixed systems and micro-dose fertilizer	Groundnut PIC bags FFS[4]
Highland tropics	Climbing bean			
Humid tropics	Disease-resistant bean and lima bean		Bean seed treatment (especially for heavy soils)	

1. Multipurpose grain legumes are designated here as 'plus', such as indeterminate cowpea (often called dual purpose). Pigeonpea local selections are abbreviated.
2. Early maturing varieties of cowpea, groundnut and bean, designated as "Early".
3. DLS = Doubled-up legume system.
4. FFS = Farmer field school. Topics include integrated crop management such as for striga, family nutrition, and market value chain linkages.

a farming system niche and identify 'best bet options' for that niche. The approach is not to find the best fit. It is to provide a diverse range of options that are biologically suited, that can be assessed through a farmer participatory process, or through performance of indicators in relationship to socio-economic criteria, to assess the fit and to adapt as needed. Gender preferences are expected to be an important aspect of a socio-ecological niche for legume crops in Africa, particularly as women often play a leading role in production and post-production processing of legumes (Ferguson, 1994).

3.1 OVERVIEW OF FARMING SYSTEM NICHES

In this section we first consider the precipitation and temperature gradients that define the major agro-ecological zones where pulses are grown. Across the arid tropics, precipitation is highly variable and less than 500 mm per annum; this context is generally not conducive to crop production and agropastoral systems dominate. There is considerable emphasis on livestock in the arid tropics, but there is some production of highly drought-tolerant pulses, such as cowpea in mixed systems with millet and with sorghum (Moussa *et al.*, 2016). Forage legumes are on occasion important in this environment, as a source of good quality livestock feed – a recent example is Burkina Faso, where farmer interest has increased regarding legume fodder. However, this is an exception for the arid tropics. As precipitation increases and becomes more reliable in the semi-arid and sub-humid tropics, food legumes become an important food crop. Most prominent among these are common bean in cool to warm environments, and cowpea, groundnut and pigeonpea in warm to hot environments. In contexts where livestock take priority, particularly in the semi-arid tropics, production of fodder may be as important as grain. Thus, cowpea and pigeonpea are widely valued as dual use or multipurpose legumes, which implies a vegetative growth type rather than grain growth types that have a harvest index.

Indeed, there is great diversity among pulse growth habits, and integration of different types is a means to enhance farming system resilience (Figure 2). Early maturing types of pulses such as 60-day common bean or cowpea provide a quick high-protein source through rapid grain production. These often have high income generation potential, as a snack or when sold as a high-protein vegetable, such as fresh bean. Others are longer-season growth types that provide multiple services, such as animal fodder, fuelwood and soil fertility, as well as some food production (Snapp, 2017). Farmers value such diversity and often deliberately grow a range of growth types. Further, farmers are often eager to experiment with new varieties so as to expand the portfolio of crop types that they cultivate. This has been shown for maize and groundnut household surveys conducted in Malawi, where farmers have often expressed interest in trying out modern varieties that were early maturing, not to replace but to augment their traditional longer-duration varieties (Fisher and Snapp, 2014).

Table 3. Pulse technologies that are 'promising options' for extension. (These legumes and management systems should be made widely available and promoted through participatory education, and policies that support access by many communities, farmer organizations and large numbers of farmers.)

Technology	Environmental fit	Socio-economic fit	Principles	Adoption	Reference
Genetic options					
Bean varieties (disease resistance)	Cool sub-humid to humid cereals and tree/tuber systems	Food insecure and market-oriented	Diversity, Farmer knowledge	Adoption of pest resistant, farmer-preferred varieties throughout Africa	Muthoni and Andrade, 2015 Sperling et al., 1993
Bean varieties (cooking time, seed quality traits)	Sub-humid to humid cereals and tree/tuber systems	Food insecure and market-oriented	Diversity, Farmer knowledge	New selection criteria for breeders; used by farmers	Cichy et al., 2015
Climbing bean, +staking	High altitude tropics, cool sub-humid	High population density, land scarce	Intensification	Adoption across all of East Africa highlands, potential for all populated highland tropics	Sperling and Munyanesa, 1995
Chickpea disease-resistant varieties	High altitude tropics	Market-oriented	Intensification	Ethiopia, other locations market constraints	Pachico, 2014
Pigeonpea multipurpose varieties	Semi-arid to Subhumid tropics	Food insecure and market-oriented	Diversity, Multipurpose (food, income, fuel, soil P, N)	United Republic of Tanzania, Malawi & Mozambique rapid growth (much less common in Central and West Africa)	Ae et al., 1990 Orr et al., 2015
Cowpea multipurpose varieties	Semi-arid	Food insecure and market-oriented	Diversity, Multipurpose (fodder, food, income)	West Africa	Alene and Manyong, 2006 Singh et al., 2003
System options					
Bean seed systems Quality declared seed (QD) systems	All	All	Diversity and farmer knowledge	3 million farmers adopted new bean varieties in Africa United Republic of Tanzania QD system	Abate et al., 2012 Pachico, 2014 Rubyogo et al., 2010
Mixtures of grain legumes (varieties grown in mixtures and legumes grown with cereals) for pest control	Sub-humid tropics	Food insecure	Diversity and farmer knowledge	Traditional farmer practices that are widespread; Research is limited as yet.	Abate et al. 2012 Ssekandi et al., 2016
Bean, cowpea cereal intercrops with fertilizer micro-dosing	Semi-arid to sub-humid	Market-oriented	Targeted intensification	Adoption with beans in some areas of United Republic of Tanzania and Zimbabwe	Snapp, Aggarwal and Chirwa, 1998 Twomlow et al., 2010
Cowpea relay intercrop with cereal	Semi-arid	Food insecure,	Diversity, Resilience	Traditional farmer practice	Kamara et al., 2011 Nederlof and Dangbégnon, 2007

Table 3 cont'd

Technology	Environmental fit	Socio-economic fit	Principles	Adoption	Reference
System options					
Seed priming	Semi-arid to sub-humid	Food insecure	Targeted intensification	Traditional farmer practice	Abdalla et al., 2015; Harris et al., 2001
DLS[1] technologies (groundnut or bean intercropped with pigeonpea) Pigeonpea ratooned	Semi-arid to sub-humid	Food insecure and market-oriented	Diversity, Multipurpose (food, income, fuel, soil P, N, soil organic matter), Intensification, Resilience	100 000s in Malawi	Chikowo et al., 2014; Snapp et al., 2010
Physical soil/water conservation (Zai pits, tied ridges)	Semi-arid, degraded soils	High population density	Targeted intensification	Traditional farmer practice, targeted use of inputs to soil and water conserved zones	Aune and Bationo, 2008; Sanginga et al., 2003
PIC bags	Semi-arid to sub-humid	Market-oriented	Targeted intensification	Contradictory reports, profitability higher on legumes but with cost risk	Baoua et al., 2012; Sudini et al., 2015

[1] DLS = Doubled-up legume system.

Multipurpose legumes

These can be described as those that combine copious vegetative growth with food products – and thus address immediate requirements for food and income along with longer-term requirements for soil building, fuelwood and livestock fodder (Orr et al., 2015). Multipurpose legumes require a long growth period to produce biomass for extra services, and so do not include short-duration growth habits. Early maturing varieties of legumes provide important services to farmers, by providing food and a source of income early in the season, and helping mitigate effects of erratic weather. These include short-duration, rapid-maturation types of groundnut, cowpea and bean that avoid drought. Mixtures of species can be used to enhance resilience through the deliberate combination of multipurpose, long-season legumes and early-maturing species (cereals or other legumes) (Figure 2; Chikowo et al., 2014;). This agroecology principle of diversification can be implemented through farmer practices, such as intercrops, polycultures and rotational sequences.

3.2 GENETIC OPTIONS
3.2.1 Warm to hot semi-arid and sub-humid tropics

Groundnut and cowpea are food legumes highly suited to smallholder farming systems in the warm to hot areas of semi-arid and sub-humid tropical savannahs (Table 2). Rapid maturing varieties that produce a crop in about 60 days have been developed for cowpea; they can be readily integrated into existing cropping systems as intercrops, relay crops or as a double-crop

> **FIGURE 2**
> **Resilience of smallholder farming systems can be enhanced through crop diversification. Legume growth habits vary from short-duration, early maturing types that produce grain in about 60 days, to long-duration, indeterminant types that produce a food crop over 6 months or more, along with other products such as high quality fodder. When grown with other crops in rotational sequences, or in mixtures, legumes lengthen the period of food production and contribute to resilience against shocks such as market variations, insect invasion and weather.**
>
> | RESILIENCE | |
> |---|---|
> | LOW | **Early maturing** |
> | | • Income |
> | | • Drought avoidance |
> | | • Food during hunger period |
> | MEDIUM | **Multipurpose** |
> | | • Food |
> | | • Soil fertility |
> | | • Fodder |
> | | • Fuel |
> | HIGH | **Mixtures** |
> | | • Buffer against pest or weather shocks |
> | | • Food throughout year |
>
> © SIEG SNAPP

sequence (Singh *et al.*, 2003). Eearly maturing groundnut varieties are also available, although there are many fewer released cultivars than for cowpea. In many cases farmers may show considerable interest in late, early or a range of maturation ecotypes, depending on the genetic material already available, and on local goals. Crops that mature at different types are often an important means to augment food security, as they either produce early when there may be limited food available, or late-duration varieties often are indeterminate in nature, and thus produce multiple harvests over a long period (Snapp, 2017). This provides a means of storage and extension of the harvest season, which is often an important constraint in dry, variable rainfall areas. However, it can also be an issue under humid conditions where storage can be problematic. In any food-insecure environments, an extended harvest of nutritionally-rich legume grain will be a welcome addition.

Groundnut is an important crop in Africa, and is particularly well adapted to the sub-humid tropics and sandy soils. West Africa is where groundnut has historically played a key role, and more recently East Africa has seen a rise in groundnut production, due primarily to expansion of area (Figure 1). Groundnut is a primary source of oil and protein in many communities, and a major source of agricultural income. However, there is a major challenge that faces groundnut production, and that is the incidence of aflatoxin – many markets have adopted what is effectively a zero-tolerance policy against this highly toxic fungal product. Aflatoxin is a critical threat to human health, and it is exacerbated by poor grain storage in hot, humid environment. A few varieties of groundnut have been developed, and plant breeding continues to address, genetic control of aflatoxin (Waliyar et al., 2016; Table 2). Currently, agronomic management and post-harvest storage are important means to achieve control (Johnson, Atherstone, and Grace, 2015). Field-tested storage solutions have been developed that not only address post-harvest aflatoxin contamination, but also protect against insect predation and poor seed germination. These include hermetic storage technologies, such as the GrainSafe Mini bag with ultra hermetic properties, and the PICS bag (Sudini et al., 2015; Williams et al., 2014). It is however important to combine these with educational efforts: thus moisture must be well-managed or such storage approaches can damage the grain. Other approaches to aflatoxin control have shown strong technical performance, and deserve to be the focus of policy support, with educational efforts, as the health hazards associated with aflatoxin require urgent attention (Wu and Khlangwiset, 2010).

Drought-tolerant, disease-resistant and farmer-approved cultivars

Over a 100 such genotypes of groundnut have been developed and released, although many of the more successful varieties were developed over 20 years ago (Table 4). Recently a number of disease-resistant varieties have been released, and are slowly becoming popular in some areas – although a lack of adoption has also been noted (Pachico, 2014), and this will be discussed later under the research section. Promising varieties include a number of disease-resistant groundnut varieties, notably to the virus-vectored disease *rosette*, which has been successfully introduced in some districts of Uganda and Kenya (Okello et al., 2014). As well, there are shorter duration varieties that have a 'bunch' growth type that requires less labour to harvest compared to traditional spreading types, as well as providing early harvest and drought avoidance, leading to widespread acceptance of two such varieties (JL24 and CG7) in Malawi (Tsusaka et al., 2016).

Another important pulse for semi-arid and sub-humid environments is pigeonpea. This crop is widely adapted to dry and variable rainfall environments; however, this species has primarily been grown to any major extent in southern and eastern Africa, where market value chains are well developed, and link to urban Asian populations and all the way to India. In

Table 4. Cultivar releases for the most important pulse crops in sub-Saharan Africa

	Pre-1970s	1970s	1980s	1990s	2000s	2010–2013	Total
Bean	1	6	22	73	130	80	312
Chickpea	0	3	2	9	12	7	33
Cowpea	3	8	49	65	32	12	169
Groundnut	20	23	25	21	48	7	144
Pigeonpea	0	0	3	2	12	4	21

Source: based on the review by Pachico (2014), updated with data from Monyo and Varshney (2016).

Ethiopia, central and western Africa, pigeonpea is grown occasionally but primarily as a home garden or border crop around a field (Table 2).

Smallholder farmers often prioritize growing maize or sorghum on most of their land, to meet food security requirements. This has proved a barrier to adoption of many pulse crops, and led some farmers to prefer legume-cereal intercrops over rotation systems. In southern and eastern Africa a pigeonpea-maize intercrop has proven to be one of the only consistently profitable and farmer-approved means to grow a pulse such as pigeonpea, which has properties that support soil conservation, while at the same time ensures a consistent supply of maize (Rusinamhodzi *et al.*, 2012). Indeed, the maize-pigeonpea system has increased in area by about 10 percent annually since 2010 in Mozambique, Malawi and the United Republic of Tanzania (Walker *et al.*, 2015). This may in part be due to rapid adoption of a medium-duration variety of pigeonpea with large stalks for fuel, and reported tolerance to flower-eating beetles (Orr *et al.*, 2015). Weevils and other insect pests remain wide-spread challenges to pigeonpea production, but there are a number of promising medium and long-duration varieties that show outstanding potential for growth in mixed cropping systems with maize, sorghum and other legumes (Monyo and Varshney, 2016; Roge *et al.*, 2016). Pigeonpea variety releases have been minimal compared with other pulse crops (Table 4), but there are a number of farmer-approved varieties of pigeonpea that deserve much greater dissemination.

3.2.2 Cool sub-humid to humid tropics

Common bean is the food legume most suited to sub-humid and humid environments, particularly in cooler areas. There is tremendous variation among varieties grown by farmers, a diversity that is further supported by the hundreds of bean varieties that have been released by governments across Africa, more so than any other legume crop (Pachico, 2014). Resistance to diseases, seed quality traits and shorter cooking time are some of the crop improvement traits that have been addressed by bean breeders in recent years (Table 3; Cichy, Wiesinger and Mendoza, 2015). These are issues of particular interest to women, who are in most cases responsible for cooking pulses, and often indicate decided preferences associated with quality traits (Isaacs *et al.*, 2016). There is considerable excitement about the fuel-saving and reduction in work load benefits associated with short cooking time that has been recently documented in eastern and central African landraces, and these traits are being incorporated into other bean lines for broader dissemination. This broadens the options available to smallholder farmers, a situation further supported by the integration of participatory action research on bean improvement with attention to supporting local seed systems (discussed in more detail below, and see David and Sperling, 1999; Rubyogo *et al.*, 2010).

Overall, as presented in Table 3, there are many varieties and cropping systems that are highly promising bean-based options ready for extension. There has been considerable progress in developing varieties and variety mixtures with enhanced disease resistance, and insect tolerance (Ssekandi et al., 2016). Providing a portfolio of varieties for farmers to experiment with in extension efforts is important, to increase genetic diversity options at the community level. Diversifying informal and formal seed options is important, as a means to support widespread availability of both commercial and non-commercial varieties. See discussion below on seed systems and technologies. Such extension should be in conjunction with support for agronomic innovations such as intercropping beans with banana, cassava and other perennials that thrive in a more humid environments (Table 3).

3.2.3 Highland tropics

Specifically adapted to the coolest zones of the tropics, in the highlands of East Africa, the climbing bean provides an example of the rapid adoption that is possible when smallholders find value in a new type of genetic resource (Sperling and Munyanesa, 1995). The associated management requirements are labour intensive, including staking or providing some form of support, yet farmers have readily learned and taken up this novel bean type over the last few decades (Tables 2 and 3). The popularity appears to be due to the tremendously high yields that are possible with this long-duration plant, as well as an ability to produce large amounts of leafy biomass. Climbing bean provides multiple products, including fresh and dry beans, as well as leaves that are edible and consumed as a vegetable. The soil improvement properties are many times greater than short-statured bean. Climbing bean is only suited to production in cold, high altitude sites, which until recently limited the production area primarily to the highland tropics of East Africa. Over 25 years, climbing bean has steadily spread throughout the highlands of Rwanda, Uganda and Kenya. An exciting new growth niche has been opened up through recent genetic improvements that have expanded the range of climbing bean adaptation to mid-altitude sites. This includes greater tolerance to heat, one of the key features of adaptation associated with these 'Mid-altitude Climbers (MAC)' (Table 4; Checa and Blair, 2012). These MAC varieties are still in the testing phase, but they are important options to consider as part of a diverse portfolio to address a changing climate, as well as for new locations.

Disease-resistant chickpea varieties have been widely adopted in Ethopia and deserve a mention here, as an option for highland and cooler areas in the tropics (Table 3). The availability and uptake of new chickpea varieties appears to have been a contributor to recognizing the increases in chickpea yield in Ethiopia; however, these varieties appear not to have been adopted elsewhere as yet (Pachico, 2014). The focus of research efforts to date have been primarily on disease-resistance and developing early maturing types with a high harvest index (Bantilan et al., 2014). As has been shown for other grain legumes, there may be need for a broad range of improved chickpea types that fit different markets and local requirements and goals (Ashby, 2009).

3.2.4 Humid tropics

Lima bean (*Phaseolous lunatus* L.) is a species that is closely related to *Phaseolous vulgaris* (common bean), and this species has potential to support bean production in hot, humid environments. This is in part due to traditional varieties of lima that have been noted to have robust disease tolerance, which is a crucial adaptation to a humid environment. Limited research investment has been carried out to improve lima bean to date, so extension efforts will need to be carefully planned to rely on local testing before promotion. It will be important to work with farmer varieties, and to consider the complex, local systems where farmers have integrated lima bean. An example of such is found in the Nigerian humid tropics, where lima is grown with African yam bean and cassava (Ibeawuchi, 2007).

3.3 MANAGEMENT OPTIONS
3.3.1 Semi-arid tropics

Mixing varieties and species is a widespread farmer practice that should not be ignored. It is an important practice that helps buffer risk, from variable rainfall, market shocks and temperature extremes as well. Further benefits from mixed cropping systems include promotion of beneficial insects, buffering of abiotic and biotic stress and, in many cases, regulation of disease. At the same time, there can be challenges associated with growing mixtures, including requirement for multiple harvest times and, often, sorting of crop produce.

Resilience to weather risk is particularly important in arid to semi-arid tropics. Planting a crop such as cowpea – which is tolerant of extremes in heat and low moisture – is a practice that acts as insurance, as it can be planted as a relay intercrop or after a cereal crop has failed due to an extreme weather event. Pigeonpea is highly suited due to phenology to planting patterns that include an intercrop with sorghum or maize, as it grows slowly initially and does not compete with the cereal. Late in the growing season pigeonpea starts to branch out, a growth pattern highly suited to erratic rainfall patterns that often see moisture becoming available after maturity of a cereal crop. Further, a pigeonpea intercrop benefits a cereal crop through biological N fixation, P solubilizing soil enhancement properties and soil organic matter conseration through leaf fall mid-season. Food production is also buffered, as harvest is distributed throughout the season in mixed plantings, as leaves become available as a vegetable early in the season with cowpea, and long-season crops such as pigeonpea provide food late in the year.

Many varieties of pigeonpea have the potential to be an important buffer against drought, through the practice of ratooning. Although this is a practice that is not part of official agronomic recommendations nor has not been systematically studied in recent decades (Kane, Rogé and Snapp, 2016), yet it is valued by some farmers, as shown by a recent survey in Malawi (Rogé *et al.*, 2017). Ratooning involves cutting branches back after the first harvest and thus harvesting two crops over time; this is a practice that should be more widely evaluated and promoted, as it saves labour and has tremendous benefits in terms of soil building and supporting beneficial insect diversity (Table 2). Consideration of the potential to ratoon should be considered as part of a pigeonpea extension programme, as an option that enhances livestock feed quality, and risk mitigation, with minimal labour requirements (however, it is important to keep in mind that free roaming livestock may curtail this practice). Overall, pigeonpea is a highly effective species at fixing nitrogen and obtains about 90 percent of its nitrogen from BNF, and has been shown to be one of the most effective crop species to enhance P availability through root exudates that enhance solubilization of P in some soil types. In contrast, bean is a very poor nitrogen-fixing crop (about 50 percent BNF rate), for the vast majority of bean types that are short-statured

Zonal soil management is another very important traditional practice, this involves targeting management of soil, water and nutrients to the zone of plant growth, and often combines modest input use with soil conservation practices such as *zäi* holes (Table 2; Aune and Bationo, 2008). However, it should be taken into consideration that a study in Burkina Faso found that farmers prioritized soil improvement through *zäi* and mulching for cereals, and did not use these practices on legume crops (Slingerland and Stork, 2000). The preferential allocation of soil amendments to cereals is an important challenge noted in many studies, and agronomic improvement of pulse production in Africa requires close attention to socio-economic context and farmer decision making. This is a barrier to sustainable production practices, as pulse production needs to be combined with soil conservation, so as to achieve soil health regeneration and protect soils. Indeed, diverse production systems that include legumes have been shown to be critical to the successful implementation of conservation agriculture.

One way forward may be to consider legume production in a systems context, as farmers invest in fertility practices that benefit legumes grown in sequence or simultaneously with cereals as intercrops (Snapp, 2017). Another form of targeted, zonal concentration of nutrients from inorganic and organic sources is the application of micro-fertilization, such as capfuls of fertilizer applied to a planting station where two or more crop species may be grown together. This approach often involves targeting of the small doses of nitrogen or P fertilizer with crop residues applied as a mulch, or complementary amendment with manure and other organic resources (Table 3; Sanginga *et al.*, 2003).

3.3.2 Sub-humid to humid tropics

Pest management becomes a pressing challenge in locations where rainfall is sufficient to support vigorous growth of weeds. Many insect and disease organisms also thrive as moisture and temperature conditions allow. Crop diversification that is promoted through intercrop systems, and planting mixtures or mosaics at the field and landscape level is one of the most important means to prevent pest damage. Regulation of pests, as well as confusing and preventing access to susceptible plants can be enhanced through cropping system design, and arrangements of different, and complementary, plantings (Snapp, 2017). Intercrops often suppress weeds, through competition for resources including the capture of sunlight. Examples include maize-beans and maize-pigeonpea, two important intercrops widely grown in the sub-humid tropics (Rao, Rego and Willey, 1987).

Diversification is important for many reasons, as it provides important buffering services for smallholder farmers that value resilience in the face of weather and market shocks. Intercropping is a historically valued practice that has been in decline and actively discouraged in some countries (Isaacs et al., 2016a, b). It has however recently begun to be re-evaluated in the literature, with key benefits documented through meta-analyses as well as comprehensive reviews (Yu et al., 2016). In an environment with sufficient rainfall, particularly bimodal systems that occur over much of the sub-humid to humid tropics, there is ample scope for intercrop and relay intercrop systems. Planting two or three crops with complementary growth types is one of the most effective means to capture sunlight and maximize utilization of water and nutrient (Kanyama-Phiri, Snapp and Minae, 1998).

Ways to enhance BNF through intercropping pigeonpea with other short-duration type legumes are considered next.

Doubled-up legume system technology relies in the initial year on the intercrop of species with complementary growth habits. In a subsequent year the system is rotated with a cereal crop that requires large amounts of nitrogen, thus the system capitalizes on the soil nutrient improvements from the cultivation of two legumes. Through diversifying species grown, the doubled-up legume system (DLS) also enhances resilience of the entire cropping system through the 'insurance' policy offered by growing crops with different requirements for moisture, and distribution of rainfall (see Box 2; Mhango et al., 2013; Smith et al., 2016).

In DLS, pigeonpea is grown as an overstory multipurpose crop that provides fuel wood, fodder, soil fertility building and food (green pods and dry grain). An understory of groundnut (or soybean, or common bean) is planted at the same time to provide an early and highly nutritious crop. As described by Mhango et al. (2013), it uses a two:one seed ratio of groundnut:pigeonpea within the same row, with pigeonpea in planting stations (two plants per station at widely spaced intervals), and groundnut as a single or double row depending on land management (double rows are usual for soil bund, or ridge or tied-ridge systems with wide spacing of rows).

3.5 STORAGE AND SEED TECHNOLOGY OPTIONS
Seed priming
In a semi-arid environment agricultural investment is often limited in nature, as variable rainfall can impose significant risk. One type of investment that imposes minimal risk and can help buffer this risky environment is that of

BOX 2.
Doubled-up legume system

Smallholder farmers across Africa grow maize, sorghum and millet in mixed cropping arrangements with food legumes. For example, sorghum is often grown in rows that are interspersed with widely spaced cowpea within the same row, or in alternate rows. Maize and groundnut or common bean are also grown as companion crops. A relatively new approach is intercropping two legumes that have different but compatible growth habits. An example is that of a pigeonpea and groundnut intercrop, where both are planted at the same time and the pigeonpea grows slowly at first, which minimizes any interference with the groundnut crop. As the groundnut begins to mature, the pigeonpea shots up and becomes a bush just as the groundnut is harvested. Thus both an understory and an overstory consist of a legume crop, providing double the food and double the biological nitrogen fixation, as well as continuous and complementary leaf coverage that protects soil from erosion. This approach – growing a legume-legume intercrop – is known as the doubled-up legume system. The Malawi government recently released this technology as an agroecology option for enhancing soil fertility while maintaining two food crops on smallholder fields. It is gaining widespread acceptance in Malawi. Farmers are experimenting and adding their own ways to manage the two crops. For example, after harvesting the double crop of groundnut and pigeonpea, some farmers 'ratoon' pigeonpea (cut back the branches) and grow it for a second year, often as a maize-pigeonpea intercrop in year two. Other farmers are adding a third crop to the mixture, such as a low growing pumpkin between rows or adding rows of maize. In whatever form farmers use doubled-up legume systems, they are incorporating the following agroecology principles into their farm:

- increasing production in a sustainable manner with double the legume presence on a plot;
- increasing agricultural biodiversity to cope with climate change;
- promoting living soil cover through crop residues that enhance soil biology, build soil nutrient status and protect against erosion.

Doubled-up legume system being used in Malawi.

seed-based technologies, such as seed-priming. A number of important seed innovations are described in Table 2. such as seed priming (preparing seeds for planting by pre-soaking them, with water and air drying, after which the seed can be stored for months). Once a seed is primed it germinates more vigorously and rapidly, for better stand establishment. The costs are minimal, making this a technology affordable to the vast majority of smallholder farmers.

Storage of legumes using PICS bags

PICS and related storage technologies produce an air-tight environment. These have a proven ability to extend storage for many months with excellent protection of seed quality. Profitability varies as there are up-front purchase costs, but profitable use is most consistently associated with PICS for higher value crops that are highly susceptible to pests, such as grain legumes (Baoua et al., 2012 ; Sudini et al., 2015).

4. How to promote pulses and legumes

4.1 SEED SYSTEMS

Support for improved function of seed systems is extremely important for crops that are not accessed through the formal seed market. This is the situation in Africa for legume crops, as about 95 percent of legume seed sown on smallholder farms is reliant on informal seed systems (McGuire and Sperling, 2016). This includes farm-saved seed, local purchase of grain to be used as seed, and kin and friend networks for sharing seed. In some cases government-sponsored access can be very important, through subsidies or seed distribution as part of disaster relief. Policy-makers and research priorities need to take into account the vital role that is played by informal seed systems, which need to be supported and improved. For example, smallholder farmers often pay for legume seed, mainly in local markets; existing channels could be used to promote new cultivars, and higher quality seed, through means such as small sized seed packets.

Over the long term, there is clear need for the creation of formal seed systems that provide smallholder farmers with access to good quality seed. Efforts to support both informal and formal seed systems should be coordinated, and both integrated with participatory plant breeding efforts to develop cultivars that are valued by farmers (Ashby, 2009). Pioneering approaches to crop improvement in West Africa have integrated farmers and farmer cooperatives from conceptualization, through variety mprovement, to developing viable seed systems (Weltzien, vom Brocke and Rattunde, 2005).

There is considerable debate on how to bolster informal seed systems. One promising approach is the enhanced production of the class of seed that is termed 'quality declared' (QD) seed, which involves local production less stringent certification mechanisms in comparison with formal, certified seed production (Abate *et al.*, 2012). This system can promote more decentralized and small-scale seed production, with lower seed prices, thus potentially increasing access to seed. To support the production of QD seed requires education and supportive polices, which have not been forthcoming in much of southern Africa. However, there are positive examples such as the widespread production of QD seed in the United Republic of Tanzania, facilitated by extension staff. It can be an effective means to promote diversity and enhance access to legume varieties, including the identification and promotion of farmer-preferred varieties.

Seed quality issues are varied, and some potentially serious problems reflect the production, handling and storage conditions. These include seed damage from pests and physical causes; the presence of seed-borne diseases; inadequate nutrition of seed; poor seed germination; and reduced vigour of seedlings. Other quality issues relate to the genetic purity of the seed, which requires attention to 'rogueing out' of off-types (plants that display a phenotype different to the variety that is being produced). This is an issue where markets require uniformity, and to address genetic purity requires understanding of how species reproduce and associated separation distances needed for production of pure seed. Legume seed quality challenges are often related to the high nutrient content of the seed, which makes pest damage and spoilage a greater concern than other crop types, as illustrated by a recent study of pigeonpea seed systems in Malawi (Jere, Orr and Simtowe, 2013).

Promoting local capacity to produce and store seed is fundamental in efforts to improve informal seed systems. This approach relies upon education and training of farmers, traders and others involved in local seed systems, to enhance understanding of how to produce, inspect and store local seed sources for improved quality (Abate *et al.*, 2012). This has been shown to be an effective means to enhance quality of local seed sources (i.e. grain that is used for seed, locally grown seed, and farm-saved seed). Examples from Kenya and Ethiopia illustrate that bean seed quality can be boosted through education of farmers and traders, as well as agricultural staff from local non-governmental organizations (NGOs). Education includes how to produce contaminant-free seed, and, in post-production, how to carry out visual inspection and hand selection to identify (apparently) disease-free, undamaged seed, with the use of simple tests to determine seed moisture content, germination and fungal presence (Odhiambo *et al.*, 2016; Oshone, Gebeyehu and Tesfaye, 2014). Related efforts include a focus on the dry season, where irrigation can be used to produce pest-free and high quality seed for the rainy season; however there has been little research reported on this approach, and evidence is required to assess this option (Kadyampakeni *et al.*, 2013).

Finally, there are efforts related to improving the quality and appropriateness of cultivars used in seed distribution as part of relief efforts. Additionally, seed fairs are used in relief efforts, to promote access to more diverse crops and genotypes (McGuire and Sperling, 2016), and fairs also feature in non-emergency contexts with NGOs or farmer organizations, as part of educational efforts to enhance appreciation of genetic diversity.

4.2 EXTENSION TO PROMOTE PULSES/LEGUMES

Extension has been shown to be more effective when client-oriented, focusing on supporting farmer experimentation and participatory approaches (Johnson, Lilja and Ashby, 2003). There are numerous opportunities to support farmer experimentation, local adaptation and adoption of technical options. This process can be promoted through education and policy initiatives. Options presented in Table 2, and in more detail in Table 3, should be considered as a basis for initiating a participatory extension and action learning process (such as a farmer field school, see Davis et al., 2012). The extension process should be an active learning approach that provides farmers and farming communities with access to options rather than imposing a set of inflexible technologies. This is an alternative to how technologies have often been promoted in Africa, where local conditions are not always taken into account (Snapp et al., 2003).

New and traditional technologies are highlighted here, with a focus on the most promising ones (Table 2). Participatory approaches to extension that involve farmer experimentation are essential for the adaptation of legumes to local environments. Collection of landraces and diverse germplasm is one way to enhance local variation in genetic resource options that can support farmer experimentation. Another is to consider a niche with few legumes. Examples include maize-dominated landscapes of southern Africa (Snapp et al., 2010), and irrigated systems (formal and informal) where paddy rice could be diversified with legumes, and legume seed production can be carried out in the low-disease-pressure dry season. New germplasm of chickpea and mung bean has been developed with adaptation to irrigated systems. This includes early and extra-early maturing varieties; these have been introduced as rotational crops to improve rice system performance in South Asia (Rashid et al., 2004).

Another very low cost, and farmer adoptable, technology is that of seed priming. It is related to traditional practices such as soaking of seeds, and selection of the best for planting, and it could be promoted in a systematic way, particularly in semi-arid areas with degraded soils, where poor stand establishment of crops is a problem. Also in areas where it could be combined with fast growing short-duration crops such as early cowpea relay intercrop (Rashid et al., 2004).

Participatory approaches can help expand the range of pulse lines so that farmers and consumers have more options. For example, participatory plant breeders use a process that often involves a wide range of stakeholders, and a commitment to releasing many different lines that meet both local and market preferences for taste and other seed traits (Witcombe et al., 2005). An important example is provided by investment in bean research at both international and national efforts (Rubyogo et al., 2010; Sperling et al., 1993), and see Table 2. This includes the sustained work of scientists, extensionists, students and others at dozens of universities and many other stakeholders who have been working together through the organized bean network 'PABRA' for over two dozen years. There appears to have been much less investment by public research institutions in other pulse crops, such as pigeonpea breeding, where few systematic efforts have been made to collect landraces, or test thousands of germplasm lines, and, indeed, a modest number of cultivars of pigoenpea have been released for African smallholder farmers. More effort is needed in this area.

4.3 EXTENSION OF CROP MANAGEMENT

Education in crop management and integrated approaches are vitally important in promotion of legumes, as legume traits that make them nutritionally rich can also pose serious production challenges (Snapp et al., 2002). This includes a modest ability to compete with weeds, at least during initial growth; being prone to insect damage (due in part to the attraction posed by nutritionally enriched plant tissues); and modest in yield relative to many crops. This requires that farmers have a plan in place to monitor for, and to control, the insects and disease organisms that often attack these crops. This requires education in agroecology, with particular attention to using intercrop systems and IPM practices. It is important to be aware that it is very difficult to grow most pulse crops in humid environments, due to pest pressure. For example, sub-humid production of cowpea is almost impossible without a means of controlling aphids and pod borers (Agunbiade et al., 2014). Thus IPM education is important for all legumes, but particularly for cowpea. Post-production utilization and storage are also challenging for pulses and require attention in a holistic manner.

Thus, integrated approaches to education, that include pre- and post-production issues such as nutrition and storage, may be key to success in promoting broader uptake of legume crops, this is represented in Table 1 as farmer field school (FFS) approaches. However, this is a broader topic of action learning. Careful attention to local priorities should drive all participatory approaches to development (Obaa, Mutimba and Semana, 2005;), and legumes may not be valued locally. This could be an opportunity for education about human nutrition, that might build local demand for legumes. However, this is an iterative, long-term process and technologies should not be introduced as a solution but rather as part of an integrated process (Neef and Neubert, 2011). In Mali, farmers have been successful in growing a wide range of cowpea and sorghum genotypes, through an integrated pre- and post-production 'farmer field school' educational effort that prioritized farmer-identified challenges such as parasitic weed control, post-harvest crop processing and market linkages. Farmer-to-farmer learning was also part of this programme, which took an integrated approach and involved extensive participation by women farmers. FFS educational approaches that involve attention to market chain integration have recently been tried out in Uganda, and show promise (Davis et al., 2012).

5. Research priorities

In this section we present an overview of research gaps, and opportunities that are critical to enhancing legume cultivation in Africa. These research priorities need to be addressed, if legumes and their symbiotic partner micro-organisms are to contribute to environmental services, and provide a foundation for sustainable agriculture.

Human nutrition
Human nutrition, and crop-livestock integration are both dependent on the protein-rich and diverse amino acid content associated with legume products, in combination with grass species. Given this pivotal role, it is startling that research investment has been so limited on tropical legume crops. This is reflected in international agricultural research budgets, as show by CGIAR core funding, which is minuscule for research on pulse crops (Pachico, 2014). Another barrier to research priority setting is the lack of information about which legumes are grown and where in global agro-ecosystems; see earlier discussion. Adoption studies for legume varieties are few and far between, and impact studies very rare indeed. Such research efforts should encouraged, and be in close conjunction with investments in participatory plant breeding and seed systems, to broaden access to farmer-preferred, improved germplasm. Finally, systematic research is lacking on ecosystem services associated with legumes.

Research priorities are considered here for a core set of legume crops, in relationship to genetics, production, post-production and context (Tables 5–9).[2] Common bean, cowpea, groundnut and pigeonpea were the primary focus, as the scope of cultivation in Africa for chickpea is primarily confined to Ethiopia, and we did not consider research priorities for this pulse. In terms of minor pulses, the literature available was limited while the research needs are many (Table 9). We suggest that researchers prioritize those genoypes that show particular adaptation to environments and provide multipurpose services, namely tropical lima bean, runner bean, African yam bean, velvet bean and hyacinth bean. There has been limited attention to documentation

---

[2] Based on literature review, these tables present topics that are a priority for research, and technologies that are close to ready for dissemination but where applied research is needed.

Table 5. Bean – research priorities for bean on smallholder farms in Africa

Technology	Long-term research	Applied research	Reference
Genetics	Yield, heat and drought tolerance Disease tolerance Nitrogen fixation enhancement Bean varieties for intercrops	Determinant, early maturity Leafy types for vegetable Climbing bean mid-altitude with heat tolerance Leafy types for vegetable Variety mixtures	Checa and Blair, 2012 Rodríguez De Luque and Creamer, 2014 Román-Avilés and Beaver, 2016 Isaacs et al., 2016a Kamfwa, Cichy and Kelly, 2015 Ssekandi et al., 2016
Production and agronomy	Conservation agriculture	Banana to provide shade with beans Staking systems for climbers Varieties suited to intercrops Sequencing studies Disease control	Amare et al., 2014 Beebe et al., 2012 Isaacs et al., 2016a TerAvest et al., 2015
Post-harvest cooking and processing		Short cooking time varieties	Cichy et al., 2015.

Table 6. Cowpea – research priorities for cowpea on smallholder farms in Africa

Technology	Long-term research	Applied research	Reference
Genetics	Insect-tolerant cowpea against polyphagous pests Insect-tolerance against monophagous pests Cowpea with large root systems Cowpea nitrogen fixation in dry areas	Determiant, early maturing type Dual use types Disease resistant types	Agunbiade et al., 2014 Kitch et al., 1998 Kristjanson et al., 2005 Sprent and Gehlot, 2010 Geleti et al., 2014
Production and agronomy	Aphid tolerance, to overcome current limits to adoption in subhumid to humid tropic	Microdose fertilizer	Buerkert and Schlecht, 2013
Post-harvest storage		Develop improved types of storage	Sudini et al., 2015
Post-harvest cooking and processing	Processing	Leafy types for vegetable Value added products	Geleti et al., 2014 Polreich, Becker and Maass, 2016

of the genetic diversity that exisits, or indeed to *in situ* conservation of this valuable germplasm resource.

Research on characterization of the diversity, and on improvements in grain quality and yield attributes, are important priorities for greater acceptance of these minor legumes. These species are often valued within traditional systems and culture but are not generally widely known, nor have crop science investments been made. Some of these species were historically of widespread importance, such as hyacinth bean, and not surprisingly, these show considerable potential for adaptation to a wide range of environments (Table 9). There are efforts underway to assess germplasm variation using molecular tools (Cullis and Kunert, 2016); however, there has been much less

Table 7. Groundnut – research priorities for groundnut on smallholder farms in Africa

Technology	Long-term research	Applied research	Reference
Genetics	Lower labour types of varieties (easy to harvest and process)	Disease resistance	Pasupuleti et al., 2013
		Nematode resistance	Tsusaka et al., 2016
	Abiotic stresses		
Production and agronomy	Sequencing and management for sustainable production and soil building	Intercrop systems	Buerkert and Schlecht, 2013
		ISFM sequencing	Harris et al., 2001
		Microdosing fertilizer	Nezomba et al., 2015
		Seed priming	Snapp et al., 2010
Postharvest storage	Genetic improvement to control aflatoxin	Aflatoxin management	Johnson, Atherstone, and Grace, 2015
		PIC and related storage systems	Waliyar et al., 2016
			Sudini et al., 2015
Post-harvest cooking/processing	Value-added products	Improved inexpensive aflatoxin screening	Dalton et al., 2012
Context (market, climate, equity)	Understanding drivers of adoption	Gender-aware studies	Ashby, 2009
		Seed systems	Pachico, 2014
			Snapp et al., 2002

Table 8 Pigeonpea – research priorities for pigeonpea on smallholder farms in Africa

Technology	Long-term research	Applied research	Reference
Genetics	Need more long duration and insect tolerant types	Pigeonpea widely adapted to different soil types and climate – semi-arid to humid	Wendt and Atemkeng 2004
			Waldman et al., 2017
Seed system Supply	Need for pigeonpea seed systems	Multipurpose fuel-food variety ready for scaling	Orr et al., 2015
			Waldman et al., 2017
Production and agronomy	Phosphorus solubilization and soil organic matter stabilization	Nitrogen and water availability rotation sequence	Ncube et al., 2009
			Snapp et al., 2010
		Doubled-up legume system (pigeonpea with understory food legume)	FAO, 2016
Post-harvest storage			
Post-harvest cooking/processing	Processing	Vegetable use	Snapp et al., 2003
Context (market, climate, equity)	Climate adaptation and risk mitigation	Female headed households valuation of pigeonpea Gender-aware studies	Snapp et al., 2010
			Mhango et al., 2013
			Smith et al., 2016

attention to phenotypic characterization within a smallholder farming context, to conservation, and landrace evaluation (Dwivedi et al. 2016). We recommend that systematic study through collections of land races and using germplasm in international collections should be a major research priority. Next would come studies of adaptation through genetic×environment assessment, and participatory on-farm experimentation will also be important.

In the next sections each pulse crop is considered in relationship to key farming system niches.

Table 9. Multipurpose and underutilized, minor legumes

Technology	Long-term research	Applied research	Reference
Genetics	Hyacinth bean Tarwi lupin African yam bean (improve grain quality, shorten cooking time, develop marketable grain)	Hyacinth bean (also known as lab lab) as a leafy vegetable Identify hyacinth bean varieties for dual use (fodder and food)	Maass et al., 2010; Geleti et al., 2014 Varshney et al., 2010
Production and agronomy	Weed suppression African yam bean and Lima bean agronomy	Doubled-up legume system (DLS) technologies Characterization of humid tropics farming systems (lima bean, yambean)	Chikowo et al., 2014 Ibeawuchi, 2007 Snapp et al., 2010
Post-harvest cooking/processing	Velvet bean and hyacinth bean processing of grain		Gilbert et al., 2004
Context (market, climate, equity)	Markets for novel legume types	Hyacinth bean adaptation to drought Gender-aware studies	Maass et al., 2010 N. Miller, pers. comm, 2016

5.1 RESEARCH PRIORITIES FOR PULSES OF THE SEMI-ARID AND SUB-HUMID TROPICS

5.1.1 Common bean (*Phaseolus vulgaris* L.)

As a crop adapted to high to mid-altitude areas, bean is widely grown across Africa and represented in almost every agroecology, to some extent. At the same time, bean yields are very responsive to temperature and rainfall patterns, and bean is a particularly vulnerable crop in a warming world, as was reflected in the priorities expressed by bean researchers and discussed below (Redden et al., 2011). Indeed, models predict that bean production area could be restricted by as much as 20 to 80 percent across Africa by 2050, unless heat tolerance is improved (Figure 3; Ramírez-Villegas and Thornton, 2015). Fortunately, progress has been made in the adaptation of climbing beans to heat and mid-altitude conditions (MAC), and there is other evidence as well that rapid progress can be made for heat tolerance in bean (Beebe et al., 2011; Román-Avilés and Beaver, 2016). Landraces of beans and close relatives provide important adaptation potential, and should be integral to research initiatives to develop beans suited to future conditions of highly variable weather and increased heat (Dwivedi et al., 2016). Longer-term research investments are required for drought tolerance than for heat; however, both are clearly needed.

Research priorities for bean system improvement were documented in a survey of African researchers, where the top priorities were related to improvements in yield under abiotic stress (drought tolerance), and improvements in grain quality (reduction in cooking time; Cichy, Wiesinger and Mendoza, 2015). The second most cited set of priorities were related to seed systems and linking farmers to markets (Rodríguez De Luque and Creamer, 2014). Other priorities included disease and insect tolerance, and notable among these was resistance to bean stem maggot. These priorities closely reflect those highlighted in a recent reviewed by plant improvement

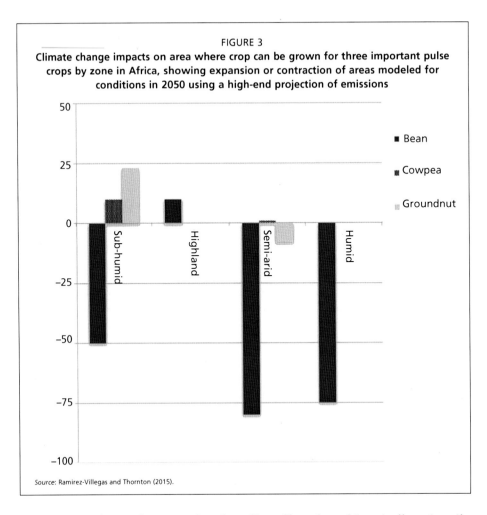

FIGURE 3
Climate change impacts on area where crop can be grown for three important pulse crops by zone in Africa, showing expansion or contraction of areas modeled for conditions in 2050 using a high-end projection of emissions

Source: Ramirez-Villegas and Thornton (2015).

scientists (Beebe et al, 2012). Plant breeding efforts have historically primarily targeted disease resistance, with many successes. It was interesting that top priorities in these two recent papers included abiotic stress and seed systems (Beebe et al., 2012; Rodríguez De Luque and Creamer, 2014). This is suggestive that new expertise and investment is urgently needed to address drought, heat tolerance and water use efficiency in beans, which are all complex traits and will require significant research investment to address.

Improvements in BNF has not been prioritized in recent bean research, despite it being an important environmental service that society seeks, particularly in maize-based ecologies, where nitrogen demands are high. The interactions of BNF with drought and other forms of abiotic stress was one additional area noted by Beebe and colleagues (2012) as requiring major, and sustained, research attention. Another research area that has been almost overlooked is that of leaf yield in bean; smallholder farmers utilize bean leaves as an important vegetable and protein source, yet this is not a trait systematically assessed by plant breeders. Rapid progress in this area could be made.

5.1.2 Cowpea [*Vigna unguiculata* (L.) Walp.]

The top priority for research in improvement of cowpea continues to be insect tolerance. Little progress has been made to date through plant breeding, and this should be urgently addressed. Research into IPM approaches is underway, including important efforts to build on traditional control measures, yet there is little evidence to date of farmer adoption of IPM, and barriers and effective extension methods need to be better understood (Tamò *et al.*, 2012). Understanding agronomic technologies that are profitable and practical to adopt is another challenging area of research, as the semi-arid and marginal lands where cowpea thrives across Africa, require judicious use of inputs and zonal management to optimize return to nutrients (Buerkert and Schlecht, 2013).

There has been rapid growth of cowpea area in recent years across West Africa, where it is an important source of protein, and in some areas, income as well (Box 1). Cowpea is growing in popularity in southern Africa as well (Figure 1). At the same time, there is little research on varieties that have been adopted, yield gaps, or understanding of barriers to adoption and performance (Pachico, 2014). One pioneering study has shown that women often express specific interest in cowpea grain quality traits, including cooking properties, storage, processing and taste, as shown among both consumers and producers in West Africa (Mishili *et al.*, 2009). This type of participatory breeding assessment through engagement with farmers, women and men, could be key to developing farmer-preferred varieties, as well as market-preferred varieties (Ashby, 2009).

5.1.3 Groundnut (*Arachis hypogaea* L.)

Research priorities in groundnut need to take into account the role of this crop as a key source of protein, oil and income across the sub-humid tropics, particularly in warmer areas. It is also a very important rotational crop with the potential to improve maize yields in a sustainable manner, and is grown as intercrop in many areas (Waddington *et al.*, 2007). Groundnuts face production challenges, but what constrains smallholder farmers the most is the lack of access to high quality seed, high labour demands associated with weed management, harvest and post-harvest, and food safety issues. Harvesting, threshing and shelling have been estimated as requiring 75 person days per tonne of groundnut (Ojiem *et al.*, 2014). A 'bunch' growth type variety in Malawi has enjoyed broad adoption in part because it is easier to harvest and to thresh, compared with traditional varieties, saving labour for women in particular (Tsusaka *et al.*, 2016), this could be considered as a trait in other breeding programmes. One major area of plant breeding improvement in groundnut has been the development of disease-resistant varieties. However, the very limited levels of adoption of varieties developed in the last two decades was highlighted in a recent review, and this suggests the need for more research attention to barriers to adoption (Pachico, 2014). This is particularly true in West Africa, where groundnut yields have generally stagnated.

Understanding drivers of adoption needs to be integrated with research priority setting, so as to enhance the development of groundnut varieties that fit a complex sets of farming system niches and consider possible gendered differences in requirements (Ortega *et al.*, 2016). Attention to gendered preferences as a key aspect of participatory plant breeding is important for successful development of pulse varieties that meet local needs, and are widely adoptable (Ceccarelli, Grando and Baum, 2007; Weltzien, vom Brocke and Rattunde, 2005). Groundnut grain in particular has a wide range of quality factors, including oil content and quality, that influence storability as well as market preference. This might require close attention in the near future if varieties are to be widely adopted (Janila *et al.*, 2013).

Maturity groups are important for the fit of a cultivar to a niche, which suggests the need for better understanding of the range of maturity types, and groundnut growth habits, this would help meet the complex preferences related to pulse traits on smallholder farms (Pachico, 2014). Farmers often are looking for a diversity in grain type as well, some grain quality traits for specific markets (condiment, oil-seed) and others with traits such as timing of production that suit needs related to food security and family nutrition. Attention to seed systems and producing a range of cultivars for market classes and home-use has been a successful strategy in bean development for African farmers, and this experience should be reviewed to determine if there are useful lessons for groundnut (Rubyogo *et al.*, 2010).

5.1.4. Pigeonpea [*Cajanus cajan* (L.) Walp.]

Pigeonpea is an important crop in specific regions of eastern and southern Africa, and the area devoted to pigeonpea has grown rapidly (>10 percent per year) in Malawi and Mozambique over the last seven years (FAOStat, accessed 20 Oct 2016). Pigeonpea has an unusual phenology, with a slow growth habit initially followed by initiation of branches from a central stem after about three or four months. As a shrub, this crop is temporarily compatible with annual crops such as maize and it is almost always grown as an intercrop (Snapp, Blackie and Donovan, 2003). Plant breeding and agronomic research has almost entirely focused on development of shorter-duration types of pigeonpea grown as sole-crops, which is a major disconnect from how pigeonpea is typically used on African farms. Research priority has rarely been given to understanding pigeonpea properties that enhance soil nutrient building, which may include root system architecture and physiological features that support biological nitrogen fixation (BNF) and P solubilization. The root exudates and rhizosphere traits, along with a long growing period, are thought to contribute to the documented ability of pigeonpea to enhance P movement from unavailable, sparingly soluble pools into plant-available pools, but how to enhance these traits has not been investigated in any detail (Myaka *et al.*, 2006). A recent study provided evidence that the role of pigeonpea rhizosphere-mediated soil aggregation may provide a crucial intermediate mechanisms that enhances P availability (Garland *et al.*, 2016), this deserves follow up.

Enhanced crop yields and stability of yields of maize sequenced with pigeonpea has been studied in field experiments across Malawi; however, there are almost no mechanistic studies reported (Snapp *et al.*, 2010). At the same time, high grain yields are rarely associated with this crop. Recent studies using selected trials have documented that pigeonpea is grown for different services depending on the farmer, where yields are a priority with some and soil fertility improvement or fuel is a priority with other farmers (Waldman *et al.*, 2017).

There is growing evidence that pigeonpea varieties are in demand that have resistance to insect pests, and that provide multiple services, as discussed above. This is illustrated by the anecdotal evidence that fuelwood, insect-tolerance and soil fertility enhancing traits are farmer-preferred attributes of cv. Mthawajumi, a non-commercial variety of pigeonpea that has spread rapidly across Mozambique and Malawi (Orr *et al.*, 2015). Farmer interviews illustrate that ratooning pigeonpea is widely practiced on smallholder farms, which supports the need for systematic research assessment of ratooning ability and association with environmental benefits (Rogé *et al.*, 2016).

Farmer interest is often centered on multiple-purpose traits, particularly in an African context. Thus it is surprising and possibly counter-productive that the goal of pigeonpea improvement has primarily focused on market traits such as large seeded type, and growth habits that are extra-short duration and erect architecture with a high harvest index (ICRISAT Happenings, December, 2015). This type of pigeonpea does not produce large stems for fuelwood, and only modest amounts of foliage for other purposes such as fodder. There is growing understanding of the need for dual-purpose types of pigeonpea, but this appears to not yet be reflected in activities or funding (Kaoneka *et al.*, 2016). A concerted effort to collect landrace pigeonpea germplasm and document performance in terms of multiple purposes, this would greatly supplement on-going crop improvement in pigeonpea. Given the observations regarding insect tolerance in a farmer-approved variety recently spreading rapidly in Mozambique and Malawi, this type of assessment of performance on-farm, and systematic documentation of farmer varieties, could lead to rapid improvement (Orr *et al.*, 2015).

5.2 RESEARCH PRIORITIES FOR PULSES FOR THE HUMID TROPICS

There are very few legumes adapted to hot, humid conditions and developing such should be a priority for future research. It will take a concerted effort, as disease and insect pressure is high under humid tropic conditions and legumes are particularly susceptible to biotic stress as discussed earlier. A few species show potential, such as tropical lima bean and African yam bean, yet we found no evidence of any systematic studies of adaptation, or efforts to improve either of these species as part of a humid tropics farming system (Table 9).

5.3 RESEARCH PRIORITIES FOR PULSES FOR THE HIGHLANDS

Crops adapted to the highlands include several *Phaseolus* species, notably runner bean, which can be grown at very high altitudes, >3 000 masl and is grown around the world in the highland agricultural niche. Many *Phaseolus vulgaris* cultivars are also suited to high altitudes, with climbing bean growth types generally grown between 2 000 and 3 200 masl. Andean lupine (*Lupinus mutabilis* Sweet) is another legume adapted to high altitudes, although its production is chiefly restricted to South America. To support broader use of this important multipurpose legume will require development of grain types that meet market demand and local preferences in Africa (Table 9). Pigeonpea can be grown up to 3 000 masl, although the cooler temperatures of high altitude sites are not conducive to high yields, and adaptation studies are urgently required in conjunction with crop improvement and participatory breeding efforts if this crop is to be promoted for highland use (Table 8; Silim *et al.*, 2007).

5.4 RESEARCH PRIORITIES FOR PULSES FOR IRRIGATED RICE-BASED SYSTEMS

Research has been very limited on legume-diversification with irrigated rice in Africa. The potential for profitable, sustainable soil management of irrigated rice has been shown in field studies exploring integration of green manure legumes and sequencing of a legume crop with a rice crop (Becker and Johnson, 1999). However, adoption has occurred only on a limited scale in South Asia (Lauren *et al.*, 2001), and no reports were found of adoption as part of Africa rice cropping. Constraints to adoption should be a major priority for research, to investigate the potential, limitations and opportunities afforded by legume diversification in African rice systems.

5.5 NEW DIRECTIONS IN RESEARCH

Legumes play an ecologically key role, due to their BNF ability and the closely related unique biochemical properties that make legume products a source of protein and other diverse nutrients (Topps, 1992). There has been limited attention to harnessing these properties in legumes.

Multipurpose types of legumes are an important option for farmers to incorporate into their farming system as they have unique properties; however this growth type has been largely neglected with the exception of fodder or agroforestry species. Food legumes can be developed, or selected from among traditional varieties, that have long growth periods that support production of copious amounts of vegetation, and deep root systems (Snapp, 2017). The leaves can be eaten as a vegetable from some species (such as bean and cowpea), and used as a fodder in almost all cases. The root system supports micro-organisms, enhances soil aggregation, and associated soil carbon accrual, and in many cases produces root exudates that solubilize P (Ae *et al.*, 1990). This builds the natural resource base and production potential of the entire cropping system. Essential to this effort will be investment in plant breeding efforts to develop a wider range of pulse genotypes with these novel traits, within farmer-preferred germplasm.

6. Conclusions and recommendations

Educators and policy-makers should consider how to promote legumes. The first step needs to be collection of better data to support understanding of where legumes are grown and for what purposes. Agricultural statistics are rather poor for pulses, including aggregated combinations of different bean species and inaccurate reflection of what is grown on the ground in many countries. Policies and protocols are needed that promote more detailed documentation of smallholder cropping systems, including tracking of many neglected legume species and varieties, as well as consumption, so that agricultural statistics would accurately reflect what pulses are grown and where. Malnutrition is an urgent problem, and plant-protein is in short supply. For more diverse, and high quality diets, education is urgently needed about the role of legumes in family nutrition.

Legumes are the basis for an ecologically-sound and farmer-focused agricultural development effort in Africa. Soil organic matter and increasing availability of nitrogen and P in a sustainable manner is fundamental for enhancing productivity around the globe. Research priorities include the need to invest in multipulses for resilient farming systems, as the focus to date on yield indicators is not sufficient to capture the multiple benefits that can be derived from inclusion of diverse legume types in farming. Weather extremes and drought require more resilient farming systems, and promoting diverse types of legume crops will help buffer farming systems. Early, fast growing types of cowpea and bean will produce grain yield even under drought conditions, and long-lived types of pigeonpea and hyacinth bean will produce grain, leaves for vegetable use, and fodder, even under highly erratic weather conditions.

7. Key recommendations

- Strengthen work on statistics reporting for pulses and other legumes.
- Promote legume innovations, such as doubled-up legume systems, climbing beans, short cooking-time beans, seed priming, and farmer-approved varieties of pulses.

- Promote farm family resilience to market and climate shocks through diversification with short- and long-duration legumes.
- Promote judicious diversification with legumes, for crop-livestock integration, and efficient and sustainable fertilizer management.
- Prioritize plant breeding that focuses on vegetative, indeterminate growth types of legumes for multiple services, pest resistance, nutritional biochemistry, and client-oriented traits.
- Support research that enhances legume reside quality and management, to improve sustainability of crop yields, through building soil carbon, nitrogen and phosphorus.
- Ensure policy recommendations take into account the significance of legumes for women and families, with attention to nutrition, food safety and processing.

References

Abate, T., Alene, A.D., Bergvinson, D., Shiferaw, B., Orr, A. & Aasfaw, S. 2012. Tropical grain legumes in Africa and South Asia: Knowledge and opportunities. Research Report, ICRISAT, Nairobi.

Abdalla, E.A., Osman, A.K., Maki, M.A., Nur, F.M., Ali, S.B. & Aune, J.B. 2015. The response of sorghum, groundnut, sesame, and cowpea to seed priming and fertilizer micro-dosing in South Kordofan State, Sudan. *Agronomy*, 5(4): 476–490.

Ae, N., Arihara, J., Okada, K., Yoshihara, T. & Johansen, C. 1990. Phosphorus uptake by pigeonpea and its role in cropping systems of the Indian subcontinent. *Science*, 248: 477–480.

Agunbiade, T.A., Coates, B.S., Datinon, B., Djouaka, R., Sun, W., Tamò, M. & Pittendrigh, B.R., 2014. Genetic differentiation among *Maruca vitrata* F. (Lepidoptera: Crambidae) populations on cultivated cowpea and wild host plants: implications for insect resistance management and biological control strategies. *PloS One*, 9(3): e92072.

Alene, A.D. & Manyong, V.M. 2006. Farmer-to-farmer technology diffusion and yield variation among adopters: the case of improved cowpea in northern Nigeria. *Agricultural Economics*, 35(2): 203–211.

Amare, A., Selvaraj, T. & Amin, M. 2014. Evaluation of various fungicides and soil solarization practices for the management of common bean anthracnose (Colletotrichum lindemuthianum) and seed yield and loss in Hararghe Highlands of Ethiopia. *Journal of Plant Breeding and Crop Science*, 6(1): 1–10.

Ashby, J. 2009. The impact of participatory plant breeding. *Pp. 649–671, in*: S. Ceccarelli, E.P. Guimaraes and E. Weltzien (eds), *Plant breeding and farmer participation*. Rome, Italy, FAO.

Asif, M., Rooney, L.W., Ali, R. & Riaz, M.N. 2013. Application and opportunities of pulses in food system: a review. *Critical reviews in food science and nutrition*, 53(11): 1168–1179.

Aune, J.B. & Bationo, A. 2008. Agricultural intensification in the Sahel – the ladder approach. *Agricultural Systems*, 98(2): 119–125.

Bantilan, M.C.S., Kumara Charyula, D., Guar, P., Moses Shyam, D. & Davis, J.S. 2014. Short duration chickpea technology: Enabling legumes revolution in Andra Pradesh India. Research Report No. 23. Patancheru, India, International Crops Research Institute for the Semi-Arid Tropics. (www.icrisat.org/what-we-do/mip/SPIA/pdf)

Baoua, I.B., Amadou, L., Margam, V. & Murdock, L.L. 2012. Comparative evaluation of six storage methods for postharvest preservation of cowpea grain. *Journal of Stored Products Research*, 49: 171–175.

Barrios, E., Buresh, R.J. & Sprent, J.I. 1996. Organic matter in soil particle size and density fractions from maize and legume cropping systems. *Soil Biology and Biochemistry*, 28(2): 185–193.

Becker, M. & Johnson, D. 1999. The role of legume fallows in intensified upland rice-based systems of West Africa. *Nutrient cycling in Agro-Ecosystems*, 53: 71–81.

Beebe, S., Ramírez, J., Jarvis, A., Rao, I.M., Mosquera, G., Bueno, J.M. & Blair, M.W. 2011. Genetic improvement of common bean and the challenges of climate change. Pp. 356–370, *in:* S.S. Yadav, R.J. Redden, J.L. Hatfield, H. Lotze-Campen and A.E. Hall (eds), *Crop adaptation to climate change.* Wiley-Blackwell.

Beebe, S.E., Rao, I.M., Mukankusi, C. & Buruchara, R.A. 2012. Improving resource use efficiency and reducing risk of common bean production in Africa, Latin America, and the Caribbean. Pp. 117–134, *in:* C.H. Hershey and P. Nate (eds), *Eco-efficiency: from vision to reality.* CIAT, Cali, Colombia.

Bekunda, M., Sanginga, N. & Woomer, P.L. 2010. Restoring Soil Fertility in sub-Sahara Africa. *Advances in Agronomy*, 108: 183–236.

Bezner-Kerr, R.B., Berti, P.R. & Shumba, L. 2011. Effects of a participatory agriculture and nutrition education project on child growth in northern Malawi. *Public Health and Nutrition*, 14(8): 1466–1472.

Blackie M. & Dixon J. 2016. Maize mixed farming systems: an engine for rural growth. Chapter in: J. Dixon, D. Garrity, J.M. Boffa, T. Williams & T. Amede, with C. Auricht, R. Lott, & G. Mburathi, (eds). *Farming Systems and Food Security in Africa: Priorities for science and policy under global change.* London and New York, USA, Routledge.

Bondeau, A., Smith, P.C., Zaehle, S., Schaphoff, S., Lucht, W., Cramer, W., Gerten, D., Lotze-Campen, H., Müller, C., Reichstein, M. & Smith, B. 2007. Modelling the role of agriculture for the 20th century global terrestrial carbon balance. *Global Change Biology*, 13: 679–706.

Buerkert, A. & Schlecht, E. 2013. Agricultural innnovations in small scale farming systems of Sugano-Sahelian West Africa: Some prerequisites for success. *Secheresse*, 24: 322–329.

Ceccarelli, S., Grando, S. & Baum, M. 2007. Participatory plant breeding in water-limited environments. *Experimental Agriculture*, 43: 411–435.

Checa, O.E., & Blair, M.W. 2012. Inheritance of Yield-Related Traits in Climbing Beans (*Phaseolus vulgaris* L.). *Crop Science* 52(5): 1998–2013. DOI: 10.2135/cropsci2011.07.0368

Chikowo, R., Zingore, S., Nyamangara, J., Bekunda, M., Messina, J. & Snapp, S.S. 2014. Approaches to reinforce crop productivity under water-limited conditions in sub-humid environments in Africa. Pp. 235–253, *in:* R. Lal, D. Mwase and F. Hansen (eds), *Sustainable intensification to advance food security and enhance climate resilience in Africa.* Springer.

Cichy, K.A., Caldas, G.V. Snapp, S.S. & Blair, M.W. 2009. QTL analysis of seed iron, zinc, and phosphorus levels in an Andean bean population. *Crop Science,* 49: 1742–1750.

Cichy, K.A., Wiesinger, J.A. & Mendoza, F.A. 2015. Genetic diversity and genome-wide association analysis of cooking time in dry bean (*Phaseolus vulgaris* L.). *Theoretical and Applied Genetics*, 128(8): 1555–1567.

Cullis, C. & Kunert, K.J. 2016. Unlocking the potential of orphan legumes. *Journal of Experimental Botany*, 68(8): 1895–1903 Special Issue S1. DOI: 10.1093/jxb/erw437

Dalton, T. K. Cardwell & T. Katsvario. 2012. External evaluation report on the peanut collaborative research support programme. Bureau of Food Security, USAID.

David, S. & Sperling, L. 1999. Improving technology delivery mechanisms: lessons from bean seed systems research in Eastern and Central Africa. *Agriculture and Human Values*, 16: 381–388.

Davis, K., Nkonya, E., Kato, E., Mekonnen, D.A., Odendo, M., Miiro, R. & Nkuba J. 2012. Impact of farmer field schools on agricultural productivity and poverty in eastern Africa. *World Development* 40(2): 402–413.

Deshpande, S.S., Sathe, S.K., Salunkhe, D.K. & Cornforth, D.P. 1982. Effects of dehulling on phytic acid, polyphenols, and enzyme inhibitors of dry beans (*Phaseolus vulgaris* L.). *Journal of Food Science,* 47(6): 1846–1850.

Dixon, R.A. & Sumner, L.W. 2003. Legume natural products: understanding and manipulating complex pathways for human and animal health. *Plant Physiology*, 131(3): 878–885.

Drinkwater, L.E. & Snapp, S.S. 2008. Nutrients in agroecosystems: Rethinking the management paradigm. *Advances in Agronomy,* 92: 163–186.

Dwivedi, S.L., Ceccarelli, S., Blair, M.W., Upadhyaya, H.D., Are, A.K. & Ortiz, R. 2016. Landrace germplasm for improving yield and abiotic stress adaptation. *Trends in plant science*, 21(1): 31–42.

Eitzinger, A., Läderach, P., Rodriguez, B., Fisher, M., Beebe, S., Sonder, K. & Schmidt, A. 2016. Assessing high-impact spots of climate change: spatial yield simulations with Decision Support System for Agrotechnology Transfer (DSSAT) model. *Mitigation and Adaptation Strategies for Global Change*. DOI: 10.1007/s11027-015-9696-2

FAO. 1994. *Definition and classification of commodities, 4. Pulses and derived products.* Rome. [Cited 22 September 2016]. http://www.fao.org/es/faodef/fdef04e.htm.

FAO. 2014. FAOSTAT: Statistics for the year 2014. Rome. [Cited 7 October 2016]. http://faostat3.fao.org/home/E.

FAO. 2016. Agroecology profile 'Integrating diverse grain legume for increased land productivity on small farms in Malawi'. Rome. [Cited 4 October 2016]. http://www.fao.org/agroecology/knowledge/practices/en/

Ferguson, A. 1994. Gendered Science: A Critique of Agricultural Development. *American Anthropologist* 96: 540–552.

Fisher, M. & Snapp, S.S. 2014. Can adoption of modern maize help smallholder farmers manage drought risk? Evidence from southern Malawi. *Experimental Agriculture*, 50: 533–548.

Fornara, D.A. & Tilman, D. 2008. Plant functional composition influences rates of soil carbon and nitrogen accumulation. *Journal of Ecology,* 96(2): 314–322.

Garland, G., Bünemann, E.K., Oberson, A., Frossard, E. & Six, J. 2016. Plant-mediated rhizospheric interactions in maize-pigeon pea intercropping enhance soil aggregation and organic phosphorus storage. *Plant and Soil*, 415(1–2): 37–55.

Gasparri, N.I., Kuemmerle, T., Meyfroidt, P., Waroux, Y. & Kreft, H., 2016. The emerging soybean production frontier in Southern Africa: Conservation challenges and the role of south–south telecouplings. *Conservation Letters*, 9: 21–31.

Geleti, D., Hailemariam, M., Mengistu, A. & Tolera, A. 2014. Characterization of elite cowpea (*Vigna unguiculata* L. Walp) accessions grown under sub-humid climatic conditions of western Oromia, Ethiopia: Herbage and crude protein yields and forage quality. *Journal of Animal Science Advances*, 4(1): 682–689.

Gilbert, R.A. 2004. Best-bet legumes for smallholder maize-based cropping systems of Malawi. Pp. 153–174, *in*: M. Eilittä, J. Mureithi, and R. Derpsch (eds), *Green manure/cover crop systems of smallholder farmers*. Dordrecht, The Netherlands, Springer.

Giller, K.E. & Cadisch, G. 1995. Future benefits from biological nitrogen fixation: An ecological approach to agriculture. *Plant and Soil*, 174: 255–277.

Glover, J. D., Reganold, J.P. & Cox, C.M. 2012. Plant Perennials to Save Africa's Soils. *Nature*, 489 (7416): 359–361.

Harris, D., Pathan, A.K., Gothkar, P., Joshi, A., Chivasa, W. & Nyamudeza, P. 2001. On-farm seed priming: using participatory methods to revive and refine a key technology. *Agricultural Systems*, 69(1): 151–164.

Ibeawuchi, I.I. 2007. Soil-chemical properties as affected by yam/cassava/landrace legumes intercropping systems in Owerri Ultisols Southeastern Nigeria. *International Journal of Soil Science*, 2: 62–68.

Isaacs, K.B, Snapp, S.S., Chung, K.R. & Waldman, K.B. 2016a. Assessing the value of diverse cropping systems under a new agricultural policy environment in Rwanda. *Food Security*, 8(3): 491–506. DOI: 10.1007/s12571-016-0582-x

Isaacs, K.B., Snapp, S.S., Kelly, J.D. & Chung, K.R. 2016b. Farmer knowledge identifies a competitive bean ideotype for maize-bean intercrop systems in Rwanda. *Agriculture & Food Security*, 5(1): 1–6.

Janila, P., Nigam, S.N., Pandey, M.K., Nagesh, P. & Varshney, R.K. 2013. Groundnut improvement: use of genetic and genomic tools. *Frontiers in Plant Science*, 4: Article 23.

Jere, P., Orr, A. & Simtowe, F. 2013. *Assessment of smallholder seed groups performance and market linkages in Southern Malawi*. Series Paper Number 12. Nairobi, Kenya ICRISAT.

Johnson, N., Atherstone, C. & Grace, D. 2015. The potential of farm-level technologies and practices to contribute to reducing consumer exposure to aflatoxins: A theory of change analysis. IFPRI Discussion Paper 01452. Washington, D.C., USA, International Food Policy Research Institute.

Johnson, N.L., Lilja, N. & Ashby, J.A. 2003. Measuring the impact of user participation in agricultural and natural resource management research. *Agricultural Systems*, 78(2): 287–306.

Kadyampakeni, D.M., Mloza-Banda, H.R. Singa, D.D., Mangisoni, J.H. Ferguson, A. & Snapp, S. 2013. Agronomic and socio-economic analysis of water management techniques for dry season cultivation of common bean in Malawi. *Irrigation Science*, 31: 537–544.

Kamara, A.Y., Tefera, H., Ewansiha, S.U., Ajeigbe, H.A., Okechukwu, R., Boukar, O. & Omoigui, L.O. 2011. Genetic gain in yield and agronomic characteristics of cowpea cultivars developed in the Sudan savannas of Nigeria over the past three decades. *Crop Science*, 51(5): 1877–1886.

Kamfwa, K., Cichy, K.A. and Kelly, J.D. 2015. Genome-wide association analysis of symbiotic nitrogen fixation in common bean. *Theoretical and Applied Genetics*, 128(10): 1999–2017.

Kane, D., Rogé, P. & Snapp, S. 2016. A systematic review of perennial staple crops literature using topic modeling and bibliometric analysis. *PLoS One*, 11: e0155788

Kanyama-Phiri, G.Y., Snapp, S.S. & Minae, S. 1998. Partnership with Malawian farmers to develop organic matter technologies. *Outlook on Agriculture* 27: 167–175.

Kaoneka, S. R., Saxena, R.K., Silim, S.N., Odeny, D.A., Ganga Rao, N.V.P.R., Shimelis, H.A., Siambi, M. & Varshney, R.K. 2016. Pigeonpea breeding in eastern and southern Africa: challenges and opportunities. *Plant Breeding*, 135: 148–154. DOI: 10.1111/pbr.12340

Kee-Tui, S.H.K., Valbuena, D., Masikati, P., Descheemaeker, K., Nyamangara, J., Claessens, L., Erenstein, O., Van Rooyen, A. & Nkomboni, D. 2015. Economic trade-offs of biomass use in crop-livestock systems: Exploring more sustainable options in semi-arid Zimbabwe. *Agricultural Systems*, 134: 48–60.

Kitch, L.W., Boukar, O., Endondo, C. & Murdock, L.L. 1998. Farmer acceptability criteria in breeding cowpea. *Experimental Agriculture*, 34(4): 475–486.

Koroma, S., Molina, P.B., Woolfrey, S., Rampa, F. & You, N. 2016. *Promoting regional trade in pulses in the Horn of Africa*. Accra, Ghana, FAO.

Koutika, L.S., Nolte, C., Yemefack, M., Ndango, R., Folefoc, D. & Weise, S. 2005. Leguminous fallows improve soil quality in south-central Cameroon as evidenced by the particulate organic matter status. *Geoderma*, 125: 343–354.

Kristjanson, P., Okike, I., Tarawali, S., Singh, B.B. & Manyong, V.M. 2005. Farmers' perceptions of benefits and factors affecting the adoption of improved dual-purpose cowpea in the dry savannas of Nigeria. *Agricultural Economics*, 32(2): 195–210.

Larochelle, C., Alwang, J., Norton, G.W., Katungi, E. & Labarta, R.A. 2015. Impacts of improved bean varieties on poverty and food security in Uganda and Rwanda. Pp. 314–337, in: T.S. Walker and J.R. Alwang (eds), *Crop improvement, adoption and impact of improved varieties in food crops in sub-Saharan Africa*. Montpellier, France, CGIAR Consortium of International Agricultural Research Centers and Wallingford, UK, CAB International.

Lauren, J.G., Shrestha, R., Sattar, M.A. & Yadav, R.L. 2001. Legumes and Diversification of the Rice-Wheat Cropping System. *Journal of Crop Production*, 3: 67–102.

Lewis, G., Schrire, B., Mackinder, B. & Lock, M. 2005. *Legumes of the world*. Kew, UK, The Royal Botanic Gardens.

Maass, B.L., Knox, M.R., Venkatesha, S.C., Angessa, T.T., Ramme, S.B. & Pengelly, C. 2010 *Lablab purpureus* – A crop lost for Africa? *Tropical Plant Biology*, 3(3): 123–135.

Marinus, W., Ronner, E., van de Ven, G.W., Kanampiu, F.K., Adjei-Nsiah, S. & Giller, K.E. 2016. What role for legumes in sustainable intensification? – Case studies in Western Kenya and Northern Ghana for PROIntensAfrica. [Cited 7 October 2016]. www.N2Africa.org.

McGuire, S. & Sperling, L. 2016. Seed systems smallholder farmers use. *Food Security*, 8: 179–195. DOI: 10.1007/s12571-015-0528-8

Mhango, W., Snapp, S.S. & Kanyama-Phiri, G.Y. 2013. Opportunities and constraints to legume diversification for sustainable cereal production on African smallholder farms. *Renewable Agriculture and Food Systems*, 28: 234–244.

MBG [Missouri Botanical Garden]. 2016. Taxonomic database Tropicos. Saint Louis. [Cited 30 September 2016]. http://www.tropicos.org.

Messina, M.J. 1999. Legumes and soybeans: overview of their nutritional profiles and health effects. *The American Journal of Clinical Nutrition*, 70: 439–450.

Mishili, F.J., Fulton, J., Shehu, M., Kushwaha, S., Marfo, K., Jamal, M., Kergna, A. & Lowenberg-DeBoer, J. 2009. Consumer preferences for quality characteristics along the cowpea value chain in Nigeria, Ghana, and Mali. *Agribusiness*, 25(1): 16–35.

Monyo, E.S. & Varshney, R.K. 2016. *Seven seasons of learning and engaging smallholder farmers in the drought-prone areas of sub-Saharan Africa and South Asia through Tropical Legumes, 2007–2014.* Patancheru, India, International Crops Research Institute for the Semi-Arid Tropics.

Moussa, B.M., Diouf, A., Abdourahamane, S.I., Axelsen, J.A., Ambouta, K.J. & Mahamane, A. 2016. Combined traditional water harvesting (Zaï) and mulching techniques increase available soil phosphorus content and millet yield. *Journal of Agricultural Science*, 8: 126–139.

Muthoni, R.A. & Andrade, R. 2015. The performance of bean improvement programmes in sub-Saharan Africa from the perspectives of varietal output and adoption. *Pp.* 148–163, *in*: T.S. Walker and J.R. Alwang (eds), *Crop improvement, adoption and impact of improved varieties in food crops in sub-Saharan Africa.* Montpellier, France, CGIAR Consortium of International Agricultural Research Centers and Wallingford, UK, CAB International.

Myaka, F.A., Sakala, W.D., Adu-Gyamfi, J.J., Kamalongo, D., Ngwira, A., Odgaard, R., Nielsen, N.E. & Høgh-Jensen, H. 2006. Yields and accumulations of N and P in farmer-managed maize-pigeonpea intercrops in semi-arid Africa. *Plant and Soil*, 285: 207–220.

NAS (National Academy of Sciences). 1979. *Tropical Legumes: Resources for the Future.* Washington, D.C., USA, National Academy of Sciences.

Ncube, B., Dimes, J.P., van Wijk, M.T., Twomlow, S.J. & Giller, K.E. 2009. Productivity and residual benefits of grain legumes to sorghum under semi-arid conditions in south-western Zimbabwe: Unravelling the effects of water and nitrogen using a simulation model. *Field Crops Research*, 110(2): 173–184.

Nederlof, E.S. & Dangbégnon, C. 2007. Lessons for farmer-oriented research: experiences from a West African soil fertility management project. *Agriculture and Human Values*, 24: 369–387.

Neef, A. & Neubert, D. 2011. Stakeholder participation in agricultural research projects: a conceptual framework for reflection and decision-making. *Agriculture and Human Values*, 28(2): 179–194.

Nezomba, H., Mtambanengwe, F., Chikowo, R. & Mapfumo, P. 2015. Sequencing integrated soil fertility management options for sustainable crop intensification by different categories of smallholder farmers in Zimbabwe. *Experimental Agriculture*, 51(1): 17–41.

Obaa, B., Mutimba, J., & Semana, A.R. 2005. Prioritizing farmer's extension needs in a publicly-funded contract system of extension: A case study from Mujono District, Uganda. *Agricultural Research and Extension Network Paper*, 147.

Odeny, D.A. 2007. The potential of pigeonpea (*Cajanus cajan* (L.) Millsp.) in Africa. *Natural Resources Forum*, 31(4): 297–305.

Odhiambo, W., Ngigi, M., Lagat, J., Binswanger, H.P. & Rubyogo, J.-C. 2016. Analysis of quality control in the informal seed sector: case of smallholder bean farmers in Bondo Sub-County, Kenya. *Journal of Economics and Sustainable Development*, 7(8): 8–29.

Ojiem, J.O., De Ridder, N., Vanlauwe, B. & Giller, K.E. 2006. Socio-ecological niche: a conceptual framework for integration of legumes in smallholder farming systems. *International Journal of Agricultural Sustainability*, 4: 79–93.

Ojiem, J.O., Franke, A.C. Vanlauwe, B. de Ridder, N. & Giller, K.E. 2014. Benefits of legume-maize rotations: Assessing the impact of diversity on the productivity of smallholders in Western Kenya. *Field Crops Research*, 168: 75–85.

Okello, D.K., Akello, L.B., Tukamuhabwa, P., Odong, T.L., Adriko, J. & Deom, C.M. 2014. Groundnut rosette disease symptoms types distribution and management of the disease in Uganda. *African Journal of Plant Science*, 8(3): 153–163.

Orr, A., Kambombo, B., Roth, C, Harris, D. & Doyle, V. 2015. Adoption of Integrated Food-Energy Systems: improved cookstoves and pigeonpea in southern Malawi. *Experimental Agriculture*, 51: 191–209.

Ortega, D.L., Waldman, K.B., Richardson, R.B., Clay, D. & Snapp, S.S. 2016. Sustainable intensification and farmer preferences for crop system attributes: Evidence from Malawi's Central and Southern regions. *World Development*, 87: 139–151.

Oshone, K., Gebeyehu, S. & Tesfaye, K. 2014. Assessment of common bean (*Phaseolus vulgaris* L.) seed quality produced under different cropping systems by smallholder farmers in eastern Ethiopia. *African Journal of Food, Agriculture, Nutrition and Development*, 14: 8566–8584.

Pachico, D. 2014. Towards appraising the impact of legume research: A synthesis of evidence. Rome Italy, Standing Panel on Impact Assessment (SPIA) and CGIAR Independent Science and Partnership Council (ISPC).

Pasupuleti, J., Nigam, S.N., Pandey, M.K., Nagesh, P. & Varshney, R.K. 2013. Groundnut improvement: Use of genetic and genomic tools. *Frontiers in Plant Science*, 4: 23.

Peoples, M.B., Brockwell, J., Herridge, D.F., Rochester, I.J., Alves, B.J.R., Urquiaga, S., Boddey, R.M., Dakora, F.D., Bhattarai, S., Maskey, S.L., Sampet, C., Rerkasem, B., Khan, D.F., Hauggaard-Nielsen, H. & Jensen, E.S. 2009. The contributions of nitrogen-fixing crop legumes to the productivity of agricultural systems. *Symbiosis*, 48: 1–17. DOI: 10.1007/BF03179980

Polreich, S., Becker, H.C. & Maass, B.L. 2016. Accession-specific effects of repeated harvesting of edible cowpea leaves on leaf yield, stability, and reliability. *International Journal of Vegetable Science*, 22(3): 295–315.

Powell, J.M., Pearson, R.A. & Hiernaux, P.H. 2004. Crop-livestock interactions in the West African drylands. *Agronomy Journal*, 96(2): 469–483.

Ramírez-Villegas, J. & Thornton, P.K. 2015. Climate change impacts on African crop production. *CCAFS Working Paper* No. 119. Copenhagen, Denmark, CGIAR Research Program on Climate Change, Agriculture and Food Security (CCAFS).

Rao, M.R., Rego, T.J. & Willey, R.W. 1987. Response of cereals to nitrogen in sole cropping and intercropping with different legumes. *Plant and Soil*, 101(2): 167–177.

Rao, V. 2000. Price heterogeneity and "real" inequality: A case study of prices and poverty in rural South India. *Review of Income and Wealth*, 46: 201–211.

Rashid, A., Harris, D., Hollington, P.A. & Rafiq, M. 2004. Improving the yield of mungbean (*Vigna radiata*) in the North West Frontier Province of Pakistan using on-farm seed priming. *Experimental Agriculture*, 40(2): 233–244.

Redden, R.J., Yadav, S.S., Hatfield, J.L., Prasanna, B.M., Vasal, S.K. & Lafarge, T. 2011. The potential of climate change adjustment in crops: A synthesis. *America*, 97: 147–152.

Richardson, A.E., Lynch, J.P., Ryan, P.R., Delhaize, E., Smith, F.A., Smith, S.E., Harvey, P.R., Ryan, M.H., Veneklaas, E.J., Lambers, H. & Oberson, A. 2011. Plant and microbial strategies to improve the phosphorus efficiency of agriculture. *Plant and Soil*, 349: 121–156.

Rockström, J., Williams, J., Daily, G., Noble, A., Matthews, N., Gordon, L., Wetterstrand, H., DeClerck, F., Shah, M., Steduto, P. & de Fraiture, C. 2016. Sustainable intensification of agriculture for human prosperity and global sustainability. *Ambio*, 46(1): 4–17. DOI: 10.1007/s13280-016-0793-6

Rodríguez De Luque, J.J. & Creamer, B. 2014. Principal constraints and trends for common bean production and commercialization; establishing priorities for future research. *Agronomía Colombiana*, 32(3): 423–431.

Rogé, P., Snapp, S., Kakwera, M.N., Mungai, L., Jambo, I. & Peter, B. 2016. Ratooning and perennial staple crops in Malawi. A review. *Agronomy for Sustainable Development*, 36(3): 50. DOI: 10.1007/s13593-016-0384-8

Román-Avilés, B. & Beaver, J.S. 2016. Inheritance of heat tolerance in common bean of Andean origin. *The Journal of Agriculture of the University of Puerto Rico*, 87: 113–121.

Rubyogo, J.C., Sperling, L., Muthoni, R. & Buruchara, R. 2010. Bean seed delivery for small farmers in sub-Saharan Africa: The power of partnerships. *Society and Natural Resources*, 23(4): 285–302.

Rusinamhodzi, L., Corbeels, M., Nyamangara, J. & Giller, K.E. 2012. Maize-grain legume intercropping is an attractive option for ecological intensification that reduces climatic risk for smallholder farmers in central Mozambique. *Field Crops Research*, 136: 12–22.

Sanginga, N., Dashiell, K.E., Diels, J., Vanlauwe, B., Lyasse, O., Carsky, R.J., Tarawali, S., Asafo-Adjei, B., Menkir, A., Schulz, S. & Singh, B.B. 2003. Sustainable resource management coupled to resilient germplasm to provide new intensive cereal-grain-legume-livestock systems in the dry savanna. *Agriculture, Ecosystems & Environment*, 100(2): 305–314.

Scherr, S.J. 1999. Soil degradation, a threat to developing-country food security by 2020? Food, Agriculture and the Environment Discussion Paper 27. Washington, D.C., USA, International Food Policy Research Institute.

Silim, S.N., Gwataa, E.T., Coeb, R. & Omanga, P.A. 2007. Response of pigeonpea genotypes of different maturity duration to temperature and photoperiod in Kenya. *African Crop Science Journal* 15: 73–81

Slingerland, M.A. & Stork, V.E. 2000. Determinants of the practice of Zaï and mulching in North Burkina Faso. *Journal of Sustainable Agriculture*, 16: 53–76.

Singh, B.B., Ajeigbe, H.A., Tarawali, S.A., Fernández-Rivera, S. & Abubakar, M. 2003. Improving the production and utilization of cowpea as food and fodder. *Field Crops Research*, 84: 169–177.

Smith, A., Snapp, S.S. Dimes, J. Gwenambira, C. & Chikowo, R. 2016. Doubled-up legume rotations improve soil fertility and maintain productivity under variable conditions in maize-based cropping systems in Malawi. *Agricultural Systems*, 145: 139–149.

Snapp, S.S. 2017. Agroecology: Principles and practice. Pp. 33–72, *in:* S.S. Snapp and B. Pound (eds), *Agricultural systems: Agroecology and rural innovation for development*. Second edition. San Diego, USA, Academic Press.

Snapp, S.S., Aggarwal, V.D., & Chirwa, R.M. 1998. Note on phosphorus and genotype enhancement of biological nitrogen fixation and productivity of maize/bean intercrops in Malawi. *Field Crops Research*, 58: 205–212.

Snapp, S.S., Blackie, M.J. & Donovan, C. 2003. Re-aligning research and extension services: Experiences from southern Africa. *Food Policy* 28: 349–363.

Snapp, S.S., Blackie, M.J., Gilbert, R.A., Bezner-Kerr, R. & Kanyama-Phiri, G.Y. 2010. Biodiversity can support a greener revolution in Africa. *Proceedings of the National Academy of Sciences of the United States of America*, 107: 20 840–20 845.

Snapp, S.S., Mafongoya, P.L. & Waddington, S. 1998. Organic matter technologies to improve nutrient cycling in smallholder cropping systems of Southern Africa. *Agriculture, Ecosystems & Environment*, 71: 187–202.

Snapp, S.S., Jones, R.B., Minja, E.M., Rusike, J. & Silim, S.N. 2003. Pigeon pea for Africa: A versatile vegetable – and more. *HortScience*, 38: 1 073–1 078.

Snapp, S.S., Rohrbach, D.D., Simtowe, F. & Freeman, H.A. 2002. Sustainable soil management options for Malawi: can smallholder farmers grow more legumes? *Agriculture Ecosystems & Environment*, 91: 159–174.

Sperling, L., & Munyanesa, S., 1995. Intensifying production among smallholder farmers: the impact of improved climbing beans in Rwanda. *African Crop Science Journal*, 3: 117–125.

Sperling, L., Loevinsohn, M.E. & Ntabomvura, B. 1993. Rethinking the farmer's role in plant breeding: Local bean experts and on-station selection in Rwanda. *Experimental Agriculture*, 29(4): 509–519.

Sprent, J.I. & Gehlot, H.S. 2010. Nodulated legumes in arid and semi-arid environments: Are they important? *Plant Ecology and Diversity*, 3(3): 211–219.

Ssekandi, W., Mulumba, J.W., Colangelo, P., Nankya, R., Fadda, C., Karungi, J., Otim, M., De Santis, P. & Jarvis, D.I. 2016. The use of common bean (*Phaseolus vulgaris*) traditional varieties and their mixtures with commercial varieties to manage bean fly (*Ophiomyia* spp.) infestations in Uganda. *Journal of Pest Science*, 89(1): 45–57.

Steele, P.E. 2011. *Southern Africa Region legumes and pulses: Appraisal of the prospects and requirements for improved food industry value addition and technical efficiency of the regional food legumes Industry. Unpublished FAO Report.* http://www.fao.org/fsnforum/sites/default/files/discussions/ contributions/FoodLegumesSouthernAfricaVersion.doc

Sudini, H., Rao, G.R., Gowda, C.L.L., Chandrika, R., Margam, V., Rathore, A. & Murdock, L.L. 2015. Purdue Improved Crop Storage (PICS) bags for safe storage of groundnuts. *Journal of Stored Products Research*, 64: 133–138.

Sumberg, J. 2002. The logic of fodder legumes in Africa. *Food Policy*, 27(3): 285–300.

Tamò, M., Srinivasan, R., Dannon, E., Agboton, C., Datinon, B., Dabire, C., Baoua, I., Ba, M., Haruna, B. & Pittendrigh. B.R. 2012. Biological control: a major component for the long-term cowpea pest management strategy. Pp. 249–259, *in:* O. Boukar, C. Coulibaly, K. Fatokun, M. Lopez and M. Tamò (eds), *Improving livelihoods in the cowpea value chain through advancements in science.* Proceedings of the 5th World Cowpea Research Conference.

Tarawali, G., Manyong, V.M., Carsky, R.J., Vissoh, P.V., Osei-Bonsu, P. & Galiba, M. 1999. Adoption of improved fallows in West Africa: lessons from mucuna and stylo case studies. *Agroforestry systems*, 47: 93–122.

TerAvest, D., Carpenter-Boggs, L., Thierfelder, C. & Reganold, J.P., 2015. Crop production and soil water management in conservation agriculture, no-till, and conventional tillage systems in Malawi. *Agriculture, Ecosystems & Environment*, 212: 285–296.

Topps, J.H. 1992. Potential, composition and use of legume shrubs and trees as fodders for livestock in the tropics. *The Journal of Agricultural Science*, 118(1): 1–8.

Tsusaka, T.W., Msere, H.W., Siambi, M., Mazvimavi, K. & Okori, P. 2016. Evolution and impacts of groundnut research and development in Malawi: An ex-post analysis. *African Journal of Agricultural Research*, 11: 139–158.

Twomlow, S., Rohrbach, D., Dimes, J., Rusike, J., Mupangwa, W., Ncube, B., Hove, L., Moyo, M., Mashingaidze, N. & Mahposa, P. 2010. Micro-dosing as a pathway to Africa's Green Revolution: Evidence from broad-scale on-farm trials. *Nutrient Cycling in Agroecosystems*, 88(1): 3–15.

Valbuena, D., Erenstein, O., Tui, S.H.K., Abdoulaye, T., Claessens, L., Duncan, A.J., Gérard, B., Rufino, M.C., Teufel, N., van Rooyen, A. & van Wijk, M.T. 2012. Conservation Agriculture in mixed crop-livestock systems: Scoping crop residue trade-offs in sub-Saharan Africa and South Asia. *Field Crops Research*, 132: 175–184.

Varshney, R.K., Glaszmann, J.C., Leung, H. & Ribaut, J.M. 2010. More genomic resources for less-studied crops. *Trends in Biotechnology*, 28(9): 452–460.

Waddington, S.R., Mekuria, M., Siziba, S. & Karigwindi, J. 2007. Long-term yield sustainability and financial returns from grain legume-maize intercrops on a sandy soil in subhumid north central Zimbabwe. *Experimental Agriculture*, 43(4): 489–503.

Waldman, K.B., Ortega, D.L., Richardson, R.B. & Snapp, S.S. 2017. Estimating demand for perennial pigeon pea in Malawi using choice experiments. *Ecological Economics*, 131: 222–230.

Waliyar, F., Kumar, K.V.K., Diallo, M., Traore, A., Mangala, U.N., Upadhyaya, H.D. & Sudini, H. 2016. Resistance to pre-harvest aflatoxin contamination in ICRISAT's groundnut mini core collection. *European Journal of Plant Pathology*, 145(4): 901–913.

Walker, T.S., Alwang, J., Alene, A., Ndjuenga, J., Labarta, R., Yigezu, Y., Diagne, A., Andrade, R., Andriatsitohaina, R.M., De Groote, H. & Mausch, K. 2015. Varietal adoption, outcomes and impact. Pp. 388–405, *in:* T.S. Walker and J. Alwang (eds), *Crop improvement, adoption and impact of improved varieties in food crops in sub-Saharan Africa*. Montpellier, France, CGIAR Consortium of International Agricultural Research Centers and Wallingforn, UK, CAB International.

Weltzien, E., vom Brocke, K., & Rattunde, H.F.W. 2005. Planning plant breeding activities with farmers. Pp. 123–152, *in:* A. Christinck, E. Weltzien, and V. Hamann (eds), *Setting breeding objectives and developing seed systems with farmers*. Weikersheim, Germany, Margraf Verlag and Wageningen, The Netherlands, CTA.

Wendt, J.W. & Atemkeng, M.F. 2004. Soybean, cowpea, groundnut, and pigeonpea response to soils, rainfall, and cropping season in the forest margins of Cameroon. *Plant and soil*, 263(1): 121–13.

Williams, S.B., Baributsa, D. & Woloshuk, C. 2014. Assessing Purdue Improved Crop Storage (PICS) bags to mitigate fungal growth and aflatoxin contamination. *Journal of Stored Products Research*, 59: 190–196.

Witcombe, J.R., K.D. Joshi, S. Gyawali, A.M. Musa, C. Johansen, D.S. Virk & Sthapit, B.R. 2005. Participatory plant breeding is better described as highly client-oriented plant breeding. I. Four indicators of client-orientation in plant breeding *Experimental Agriculture*, 41: 299–319.

Wu, F. & Khlangwiset, P. 2010. Evaluating the technical feasibility of aflatoxin risk reduction strategies in Africa. *Food Additives & Contaminants: Part A*, 27: 658–676.

Yu, Y., Stomph, T.J., Makowski, D., Zhang, L. & van der Werf, W. 2016. A meta-analysis of relative crop yields in cereal/legume mixtures suggests options for management. *Field Crops Research*, 198: 269–279.